Maritime Archaeology in Australia

Divers mount the harpoon platform on a recently scuttled whale-chaser. (Photo: Pat Baker)

GRAEME HENDERSON

Maritime Archaeology in Australia

 UNIVERSITY OF WESTERN AUSTRALIA PRESS 1986

First published in 1986
by University of Western Australia Press
Nedlands, Western Australia

Agents: Eastern States of Australia, New Zealand and Papua New Guinea: Melbourne University Press, Carlton South, Vic. 3053; U.K., Europe, Africa and Middle East: Peter Moore, P.O. Box 66, 200a Perne Road, Cambridge CB1 3PD, England; U.S.A., Canada and the Caribbean: International Specialized Book Services Inc., P.O. Box 1632, Beaverton, Oregon 97075, U.S.A.

This book is copyright. Apart from any fair dealing for the purposes of private study, research, criticism or review, as permitted under the Copyright Act, no part may be reproduced by any process without written permission. Enquiries should be made to the publisher.

© Graeme Henderson 1986

National Library of Australia
Cataloguing-in-Publication data

Henderson, Graeme, 1947- .
    Maritime archaeology in Australia.

    Bibliography.
    Includes index.
    ISBN 0 85564 241 6.

    1. Underwater archaeology—Australia. 2. Underwater archaeology—Australia—History. 3. Shipwrecks—Australia. I. Title.

994

Photoset by University of Western Australia Press, Nedlands, Western Australia.
Printed and bound by Globe Press Pty Ltd, Brunswick, Victoria.

Contents

	List of Illustrations	vi
	Acknowledgements	viii
1.	Introduction	1
2.	What is Maritime Archaeology?	5
3.	Ships in Australian Waters Before European Settlement	16
4.	Ships in Australian Waters After European Settlement	33
5.	Finders and the Law	67
6.	Early East Indiaman Traders: Pre-Settlement Maritime Archaeology at the Western Australian Museum	79
7.	Post-Settlement Maritime Archaeology at the Western Australian Museum	101
8.	Adventure and Misadventure in the South Pacific: Maritime Archaeology at the Queensland Museum	128
9.	Maritime Archaeology in Tasmania, Victoria, South Australia, New South Wales and the Australian Territories	145
10.	A Reflective Overview and Prospects for the Future	166
	Appendix: Shipwrecks Declared Under Legislation	173
	Bibliography	179
	General Index	187
	Index of Vessels	199

List of Illustrations

Divers mount the harpoon platform on a recently scuttled whale-chaser	*Frontispiece*
The Australian continent	13
The Cape of Good Hope	17
A model of the VOC ship *Zuytdorp*	25
One of the guns from Napier Broome Bay	29
Captain John Hunter	35
Admiralty draught of the *Sirius*	35
A model of HMS *Pandora*	39
Carved slate, thought to represent the *Sydney Cove*	43
Sailing vessel rigs	46
A representation of the *Edwin Fox* as a convict transport	49
The *Edwin Fox* as a hulk	50
Barrels of provisions lie stacked on the wreck of the *William Salthouse*	53
A section of the *William Salthouse* site plan	54
Barrel lids from the *William Salthouse*	54
Sites gazetted under the Commonwealth Historic Shipwrecks Act	56
The clipper ship *Schomberg*	57
Passengers at Fremantle	61
The pearling lugger *EWS*	63
The Aberdeen White Star Line's 11,400-ton TSS *Pericles* going down	64
The battered German light cruiser *Emden*	65
The Western Australian Maritime Museum	74
The remains of one of the castaways from the *Batavia*	81
The 11-metre Museum workboat *Henrietta*	83
An unexpected large wave overwhelms the *Henrietta*	83
The *Henrietta* emerges from the wave	83
The Dutch fort at Batavia	85
Sandstone blocks raised from the *Batavia*	86
The port side of the transom of the *Batavia*	88
Cleaning ballast bricks from the *Vergulde Draeck*	92

LIST OF ILLUSTRATIONS

Clay smoker's pipes from the *Vergulde Draeck*	93
Collection of *Zeewijk* (1722) material	95
The turbulent *Zuytdorp* wreck site	97
The small sand cay on Clerke Atoll	103
Access to a sheltered boat anchorage, Clerke Atoll	104
Plan of the *Rapid* site	105
Recording the profile across the *Rapid* wreck	106
Preparing parts of a powder canister for packing	107
Heavy ballast is removed from the *Rapid* site	108
An American ship in distress	111
Chess pieces from the *James Matthews* site	115
Measuring the hull and slate cargo of the *James Matthews*	117
Checking a pintle lying on the Rowley Shoals shipwreck site	119
A shipwreck site not protected by legislation	121
A cannon is lifted from the shallow *Cumberland* wreck.	122
The London Times announces the departure of the *Eglinton*	123
An isometric sketch of the SS *Xantho* wreck	125
De-concreting the steam engine from the wreck of the SS *Xantho*	125
Cheynes 3 is sunk with explosives in shallow water	126
The scuttled whale-chaser *Cheynes 3*	126
Turtles lay their eggs on a group of sand cays	130
The rudder pintle from HMS *Pandora*	132
A photomosaic of the *Pandora* site	137
A remote-piloted vehicle films divers using a water dredge	140
A cabin fire-place, chimney intact, is exposed on the *Pandora* wreck	141
Ceramic amulets from the wreck of the *Foam*	143
The landing craft used in a feasibility study of the *Sydney Cove* wreck	147
Diagram showing comparative keelson construction methods	149
The frames of the heavily built *Rapid* are packed tightly together	149
A sand glass from the *Sydney Cove* wreck site	152
Ceramic plate, bowl and jar from the *Sydney Cove*	152
Museum visitors by the helm of the *Polly Woodside*	155
The hull of the *Miranda* lies buried on the beach at Wilson's Promontory	160
Turbulence clouds the seabed at the *Sirius* wreck site	163
Rudder chains from the *Sirius* being sketched	164
Lamp black being used as a restoration treatment on a cannon	164

Acknowledgements

When I started to write this book back in 1976 it was an unattainable challenge. Maritime archaeology had been started in one Australian State, but was not practised anywhere else in this country, so one could not discuss 'maritime archaeology in Australia'. Thus the book lapsed. It has now been finished because this field has blossomed, and is actively pursued throughout Australia. That progress is a tribute to community co-operation, to what the American archaeologist Charles McGimsey has referred to as 'public archaeology'. Maritime archaeology in Australia has its greatest strength in grass-roots level support.

I could not have attempted to record the developments taking place in most Australian States but for the support given by the Director (John Bannister) and Trustees of the Western Australian Museum, who allowed me to direct and participate in crucial fieldwork throughout Australia. My understanding of these developments was also assisted by support in particular projects from the Australian Research Grants Scheme, the Literature Board, the Department of Arts, Heritage and Environment, the Australian Bicentennial Authority, the Queensland Museum and the National Parks and Wildlife Service of Tasmania.

I have to thank my colleagues from various Australian States who helped me with sections relating to their particular region: in particular Bill Jeffery of the Department of Environment and Planning in South Australia, Mark Staniforth and Peter Harvey of the Victoria Archaeological Survey, Peter Gesner of the Queensland Museum, Paul Clark of the National Parks and Wildlife Service of Tasmania, and Michael Lorimer of New South Wales.

My colleagues from the Western Australian Museum gave much assistance and advice. Patrick Baker helped with the illustrations and Myra Stanbury, Sally May and Kate Morse made many useful general comments. Dr Ian Crawford (who, as the Head of the Division of Human Studies, was responsible for the creation of a department of maritime archaeology), gave me the value of his insight, and Dr Ian MacLeod read the sections dealing with conservation.

Others gave me the benefit of their specialist knowledge. Dr Patrick O'Keefe, Senior Lecturer in Law at the University of Sydney, read through the legal section and gave valuable advice. I am particularly indebted to Associate Professor Frank Broeze, a specialist in Maritime History at the University of Western Australia, who read through the drafts several times and gave many valuable suggestions. Professor

ACKNOWLEDGEMENTS

Sandra Bowdler, Head of the Department of Prehistory at the same university, and the proponent of the coastal colonisation model for Australian prehistory, gave valuable comments. Peter Ryan of the Commonwealth Department of Administrative Services provided information about the operation of the Commonwealth's legislation. Mike McIntyre, Director of the Victoria Archaeological Survey, commented on the Victorian section.

James Henderson commented on the drafts as a writer. Kandy-Jane Henderson convinced me that I should complete the book.

Among the many names of people I recall who in early years helped in diverse ways in the genesis of the field of maritime archaeology in Australia are Dr Phillip Playford, John Cowan, Dr David Ride, Mike Pollard, Max Cramer, Professor Geoffrey Bolton, Hugh Edwards and Eric Christiansen. I was initially inspired to write the book after reading the work of Professor John Mulvaney, an historian turned archaeologist. Vic Greaves and Eric Wood encouraged me from the inception of the project. It was a pleasure to work with the staff of the University of Western Australia Press.

The drafts were typed by the dextrous fingers of Cynthia Baker, Sue Cox and Lucy Marchesani.

The views expressed in the book are my own and not necessarily those of the Trustees of the Western Australian Museum.

1 Introduction

An enormous volume of archaeological material lies, undisturbed by man, on the Australian seabed. The material representing Australia's historical period consists mainly of shipwreck sites. Shipwrecks generally occurred as a result of a disaster of one sort or another, in dangerous waters, and were frequently for these reasons immediately abandoned and left alone by survivors and succeeding generations. The water itself provides a good medium for preservation, so some shipwrecks have remained relatively intact. It would be difficult today to mount a case in favour of archaeologists, historians and anthropologists ignoring these sites as a cultural resource of value in answering questions about the past, and of human behaviour.

Nevertheless, 25 years ago all the archaeologists working in Australia, whether they were prehistorians or historical archaeologists, regarded archaeology as something you did on land. The important questions about desert-dwelling nomads were to be answered in the dry interior, and questions about colonial settlers were to be answered among ruins in the old cities and towns, or on the early homesteads. Old cultural items found on the seabed were not of any interest to these archaeologists. Nor, for that matter, were archaeologists in other parts of the world interested in seafarers.

Now the balance in Australia has been redressed. 'Land' archaeologists have recognised the importance of the sea to past generations of people, and some archaeologists have ventured into the water. Like mainstream archaeology and history, maritime archaeology must henceforth be recognised as being useful within our society for the perspective it gives, the alternatives it demonstrates can exist, and the wisdom it allows people to gain.[1] Maritime archaeology, although still a very small field when compared with Australian prehistory, which has over 200 practitioners, is practised keenly in Australia. Ten maritime archaeologists are employed in permanent positions in institutions representing all States except New South Wales, while temporary and support staff number about as many again. The membership of amateur maritime archaeological groups totals many hundreds. Why has the field blossomed here, when many other countries with older remains lying underwater continue to take the attitude that archaeology stops at the high water mark?[2]

There are several reasons for the successful train of events in Australia. In the beginning, in 1963, the first important shipwrecks were found by concerned citizens—divers with a sense of responsibility towards what they saw as a part of Australia's history. By chance, the finders were closely associated with interested journalists. So from the outset two necessary ingredients for the beginnings of maritime archaeology were present—a grassroots pressure group combined with media support. In addition the economy was growing, and a State institution (the Western Australian Museum) was prepared to accept responsibility for historic shipwrecks. Old shipwrecks and seafaring were not generally considered valid areas of study among academic archaeologists and anthropologists but the Western Australian Museum could in its more detached situation develop maritime archaeology.

The answer also lies partly in the very fact that European culture on this continent has only a short history. The seventeenth century Dutch and English shipwrecks, for example the *Trial*, pre-date the first European settlement by up to 166 years. When the Australian bicentennial year celebrations are held in 1988 the English *Trial* wreck will have been lying in Australian waters for 366 years. Although the East Indiamen shipwrecks are not a direct part of our cultural heritage, their excavation has given Australians a sense of association with an exciting and seemingly more glamorous past. Maritime archaeology has been established in Australia as a result of the discovery of the wrecks of the seventeenth century Dutchmen.

A sense of high adventure was maintained in the minds of both participants and observers of maritime archaeology throughout the first decade of its practice in Australia. The first sites investigated lay in inaccessible areas, on top of storm-battered coral reefs where calm water was never experienced. But that was no deterrent. Rather, the physical obstacles of the environment were a positive stimulus, seen as part of the challenge of proving that archaeological methods could be used to draw the information from these underwater time capsules, being as they are a closed or discrete group of cultural finds from a particular point in time. It seems appropriate that pioneering work in this new field of research should have been carried out in a young, vigorous, resource-rich nation such as Australia.

Now that maritime archaeology has been established in Australia, the emphasis of the work has shifted away from the Dutch shipwrecks to reveal a new range of facets of colonial and maritime history and technology through the investigation of post-European-settlement period shipwrecks. This has occurred not because excavations have been completed on the earlier sites. Indeed the earliest wreck in Australian waters has until very recently not received any archaeological attention beyond a brief survey in 1971. Rather, there is a growing interest in shipwreck sites which can provide information directly relating to the people who have settled and developed the continent. As the nation matures and becomes more multicultural Australians both new and old want to find a national identity to give themselves real roots here.

Ships have been wrecked on every section of the Australian coastline. These shipwrecks are almost daily being discovered by a growing army of sports divers who have the equipment, mobility and enthusiasm to explore all the shallower areas of the continental shelf. In States where funded maritime archaeologists are operating, curious divers have the opportunity to see at close hand the archaeological aspects of some of these sites and, if they are interested, can participate in expeditions by joining one of the maritime archaeological associations. Most of the shipwreck sites excavated in Australian waters to date have been located in difficult environments, where manpower requirements are heavy, and volunteers an essential expedition component. A typical staff-to-volunteer ratio on a Western Australian Museum archaeological expedition, for instance, is 1:5. This does not mean that all the volunteers record and excavate underwater. Staff and volunteers operate as their particular skills best suit them, and take part in the equally important tasks of artefact registration and sorting, illustration of finds, boat-deck duties, equipment maintenance and camp-kitchen duties.

Within Australia, immense interest has been shown in maritime archaeology by most sections of the community: divers, the museum visiting public, teachers and students, journalists, amateur and professional historians and archaeologists, scientists, politicians, and even lawyers. The interest is not restricted to men. An increasing number of women are being attracted to the field, and they participate at all levels. I have tried to respond to the needs of the general reader, but the specialist too will gain useful insights because this, the first book on the subject, is written by a professional insider.

This book is not a handbook of techniques for maritime archaeology. Several of these have already been written.[3] Nor is it an analysis of the theory of maritime archaeology, although much is needed on that subject. Maritime archaeology in Australia is still in the pioneer phase—there is a substantial published data base (or set of interpretations) but a theory of maritime archaeology has yet to be developed. The purpose of this book is to provide for the first time a comprehensive account of the unique origins and development of maritime archaeology as a field of study in Australia, and to assess its current position. There are good reasons for doing this.

It is important that the information be made available to the Australian public—particularly as the field is so congenial to public involvement. Maritime archaeology in Australia owes a great deal to concerned private citizens. Pressure groups of amateur archaeologists (principally skindivers) have been successful in persuading governments to bring comprehensive protective legislation into effect. Coupled with the legislation, sympathetic government institutions have initiated State programmes with strong support from amateur archaeological organisations. These groups must be kept informed.

Overseas journals contain many articles by Australian maritime archaeologists describing and analysing individual sites in Australian waters. Communication of broader issues, however, has been severely limited. The distance between Australia and the Atlantic countries has been sufficient to limit the information flow. In particular, few maritime archaeologists working in Europe and America have visited sites and institutions in Australia. Furthermore, no overview has been available to give Atlantic scholars a perspective view of the developments of the past 20 years, and the expansion of the field to all States. There are books, recently published, on Australian prehistory and historical archaeology, but these give no hint that maritime archaeology exists, let alone flourishes, in Australia.

The following chapters attempt to provide a critical overview of the achievements and current state of maritime archaeology in Australia. The first section addresses the question—what is maritime archaeology? Then the activities of ships in Australian waters are discussed, with comment on the value of particular types of shipwrecks as a potential resource for maritime archaeologists. Misinformation has been circulated regarding the rights of finders of shipwrecks, and the implications of some clauses of the protective legislation and the outcomes of several court cases dealing with historic shipwrecks. The chapter 'Finders and the Law' is intended to clarify some of these issues. The beginnings and growth of the field in Western Australia and all Australian States are then outlined in some detail. The final chapter reflects upon the prospects for the future. The deficiencies of the book are a reflection not only of my own limitations, but also to some extent those of the profession.

Readers will notice a strong emphasis upon fieldwork and legislation in the following pages. This is a reflection of the central role played in the genesis of Australian maritime archaeology by the Western Australian Museum (which like any museum sees the acquisition of collections as a part of its function) and of the crucial role of the Commonwealth legislation in the rapid spread of maritime archaeology from Western Australia to all other Australian States.

Chapter 1   REFERENCES

1. Partial quote from Peter White and O'Connell, 1982, p. 3.
2. See O'Keefe and Prott for a review of worldwide legal controls.
3. For example, Bass, 1966, UNESCO, 1972.

2 What is Maritime Archaeology?

The purpose of this chapter is to define the area of study and survey the sorts of sites which might prove useful in shedding light on questions asked by underwater archaeologists and maritime archaeologists. So I have cast my net widely to include all archaeological sites underwater, whether or not they are maritime, and all maritime archaeological sites, whether or not they are underwater.

First, I shall explain what is meant by 'maritime archaeology' (the term gained its popularity from the Western Australian Museum's use of it following the Maritime Archaeology Act of 1973). It is a relatively new area of study, and has only recently reached a position where the data base is sufficiently extensive to allow some tentative steps in defining it as a branch or sub-discipline of archaeology. Two archaeologists who have considered the question at length after working on underwater excavations are George Bass and the late Keith Muckelroy.

Bass was the first professional archaeologist to carry out a major excavation entirely underwater. An American, he received his formal academic training in classical studies and classical archaeology, and as a doctoral candidate directed underwater excavations in the Mediterranean on a Bronze Age shipwreck at Cape Gelidonya and a Byzantine shipwreck at Yassi Ada. He pioneered the scientific approach to excavation underwater. His book *Archaeology Underwater* was published in 1966, a time when he was arguing, quite justifiably, that excavations carried out underwater should be recognised as being archaeology.[1] He was trying to convince a sceptical public, which had previously only seen shipwrecks as the realm of treasure hunters, that archaeologists who jump into the water to look at these sites are archaeologists notwithstanding. Archaeology underwater should be called simply 'archaeology', he argued. We don't talk of mountain archaeologists or jungle archaeologists, and there is no need to speak of underwater archaeologists as being people with some different aim—it is all archaeology. Bass's book was concerned primarily with the techniques necessary for doing archaeology underwater, rather than with the development of a theory of maritime archaeology. He included in his concept of 'archaeology underwater' just that: everything underwater, whether it be in oceans, lakes, rivers, streams, swamps, or even wells. The thing was that you had to get wet. So he excluded all dry sites, even such things as shipwrecks on drained land.

Different sorts of archaeologists are going to be interested in shipwrecks, submerged cities, prehistoric migration patterns, and well sites. They may have little in common beyond getting wet and having to adapt their excavating techniques accordingly. The subject areas and the problems being investigated will be widely divergent.

Bass's 1966 concept of 'archaeology underwater' lacks any cohesion (for people wishing to define a new department or sub-branch of knowledge) except in terms of adaptation of techniques to suit the underwater environment. There can be no theoretical framework for 'archaeology underwater'. The title of his 1972 publication—*A History of Seafaring*—is much more useful in this respect. Nevertheless, Bass's earlier concept of archaeology underwater, his classical background, and the fact that he did his pioneering work on bronze-age sites in very deep water, has had strong implications for the development of the field—there has been a particular emphasis on the developing of special equipment for operating in the environment. Bass himself clearly regrets that perceptions of what he was doing affected the field in this direction. He writes:

> The techniques were not an end to themselves, but only aimed for greater efficiency in underwater work.
>
> Because the same archaeologists (some time later) were still analysing their sites through research and publishing only brief preliminary reports, it must have seemed to the public and to other archaeologists that underwater archaeology was mainly a technological field filled with diving bells, submarines, sonar, underwater decompression chambers, underwater television . . .
>
> Archaeologists tended to emphasise their futuristic gadgetry, sometimes because funding of their work came from navies or foundations devoted to applied science rather than to the humanities . . .[2]

A decade after Bass's *Archaeology Underwater* Muckelroy wrote his *Maritime Archaeology*, which draws together the variety of work that had been going on in various parts of the world in the 1960s and 70s; further emphasises Bass's point that seafaring was an essential part of man's progress—culturally, socially, economically and technologically; and shows the influence of the department of archaeology of Cambridge University[3]. In this he attempted to define a new sub-discipline: maritime archaeology is 'the scientific study of the material remains of man and his activities on the sea'.[4]

The emphasis was on the term 'maritime', that is, anything connected with seafaring in its broadest sense. Muckelroy's sites could be situated underwater or on land. He included ships, boats, and their cargoes, along with harbours, but excluded sunken cities, lake sites, river, stream, and well sites, and sea-fronting material. He included all aspects of maritime culture: not just technical matters but also social, economic, political, religious etc.

His definition was intended to isolate those aspects in which ancient ships and their cargoes, along with toppled harbour walls, are different from non-maritime material, and so identify a new archaeological sub-discipline. This was a further move towards problem oriented research, and development of theory.

The concept of the archaeology of seafarers has some theoretical cohesion in the same way that frontier archaeology, industrial archaeology and Islamic archaeology have. And the dominance of underwater sites has resulted in the creation of a set of techniques distinctive to maritime archaeology. So Muckelroy's claim to have described a new sub-discipline has some justification.

In the immediate future at least, I have no doubt that most archaeologists working in the underwater environment in Australia will think of themselves as underwater archaeologists first, and as maritime archaeologists only secondarily. But the concept of 'maritime' is visible in the titles of legislation, amateur and professional associations and institutions.

Purely practical cost considerations press archaeologists into specialising in particular types of sites and a finite time span. The same considerations influence the structure and development of a department of archaeology with its laboratory plant and field equipment. So an individual department of archaeology once established cannot freely alternate between terrestrial prehistory, early postmedieval maritime archaeology and modern industrial archaeology. It would require costly changes in equipment and personnel. Few institutions in today's world are likely to invest the sums necessary to carry out a maritime archaeological excavation without contemplating a permanent commitment. These financial considerations are as important as the more academic questions about a cohesive subject area: an archaeological institution with a heavy investment in work boats is more likely to pay attention to archaeological sites underwater, whether they be shipwrecks or sunken cities. The structure of an involved institution influences its approach to maritime archaeology. A small minority of institutions are large enough to combine the roles of site management, conservation, curation, display etc. under the same roof, while smaller institutions are restricted to site management, and must attempt to gain access to separate institutions which engage in the other roles. In these smaller institutions great emphasis is placed on 'cultural resource management', a phrase coined to describe a development in United States practice whereby all federal government agencies are required to conduct preliminary surveys, identify significant sites and take measures to minimise damage.

A recent American publication, *Shipwreck Anthropology,* may well stimulate further debate about the direction of maritime archaeology. Most of the contributors to the volume are anthropologists, and they argue in favour of an anthropological approach to the archaeology of shipwrecks. The anthropologists ask what a shipwreck archaeologist can do to discover general principles that hold true for both past and present-day human nautical behaviour, and they emphasise that what makes shipwreck archaeology a science is not the use of scientific techniques and equipment but scientific method—the testing of alternative hypotheses, parsimonious use of resources, the need for repeatability of results, and the ability to extend the results from a particular case to the realm of general considerations about the nature of human behaviour. Some shipwreck archaeologists have made systematic efforts to control for a wide array of natural variables such as currents, sea-bottom conditions, salinity, and other factors in explaining the particular characteristics of different kinds of wrecks. But comparable controls in explaining how behavioural variables operate to produce different kinds of physical associations are still uncertain and untried.[5]

The anthropological approach, with its emphasis on culture process (the analysis of change) rather than culture history (lifeways), is referred to as 'new' archaeology. It emphasises the need for explicit theory and rigorous methodology. From an early stage American archaeology was inextricably wed to anthropology through use of living people viewed as an analogy with which to interpret the artefacts of the past.[6] Australian prehistoric archaeologists have welcomed this new approach. No one knows exactly how the early Aborigines lived or exactly what behaviour their archaeological record represents. So Australian prehistory must by its nature be a predictive science, relying heavily on the use of general principles or laws about human behaviour, and a major focus has been to develop more sophisticated methods of prediction.

Prehistory deals with the period prior to written records. It relies very much therefore on archaeological methods in conjunction with understanding of physical processes, such as those studied in geology, geomorphology, botany etc. This makes it a particularly dynamic study, with interpretations changing sometimes drastically as new information is acquired.

Historical archaeology is to some extent a different type of archaeology from prehistory, because it utilises written records and does not need to depend so much on anthropological theory and the use of analogy. However, historical archaeologists today are increasingly aware of the value of these latter methods. 'Archaeological' problems (whose answers will further the development of predictive patterns) are legitimate because the written evidence is often lacking or misleading. But it should also be acceptable for historical archaeologists to address 'historical' problems, whose answers provide insights about past events or processes.[7]

Most historians think that human behaviour cannot be predicted absolutely by general laws. They believe in the operation of free will and think that while human behaviour exhibits patterns and regularities, there are nevertheless always exceptions. Anthropologists would agree, but tend to place greater emphasis on the seeking of regularities.

Australian maritime archaeology currently deals almost entirely with historical period sites, so its standpoint might be expected to parallel that of historical archaeology. However, the implications of the 'new' archaeology have been slower in their influence on maritime archaeology in Australia.

Maritime archaeology as practised in Australia has interdisciplinary usefulness— in particular it is making a contribution to the study of maritime history, Australian history and European history. Information relating to many aspects of maritime history—ship construction and fitting out, arming of ships, navigation techniques, the way of life of crew and passengers—is difficult to glean from available documentary sources. Maritime archaeology provides the means for testing some long held theories in maritime history, and provides material for establishing new theories. The cargoes lying in the holds of wrecked ships provide a resource of information about many aspects of Australian and European history such as, for example, trading practices and commerce.

Public maritime archaeologists (working in museums and other government heritage management institutions and frequently employing contract archaeologists) carry out activities directly associated with 'monuments' (in this case generally substantial ship's structure) and archaeological sites, such as inspecting newly found sites, preparing management plans and proposals for new or improved legislation, surveying and excavating sites, researching archaeological collections, and generally curating (sometimes conserving) those collections. But they are also inevitably involved in dealing more directly with the general public's information needs. This involves talking to various groups as part of a public education programme; answering questions from private laymen, journalists, other institutions and industry about shipwrecks; identifying artefacts in private custody; preparing news releases on work carried out; and assisting in the formulation of displays of parts of the collection. The spectacular appearance of some of the maritime collections, for example the remains associated with the *Batavia* massacre, and the medical instruments from HMS *Pandora*, has given them strong appeal with the general public, so maritime archaeology has become perhaps the most obvious branch of archaeology in the media and public exhibition venues. The high public profile presents some disadvantages however. Diving and the raising of spectacular artefacts are photogenic, but library and museum research are not.

The scope of maritime archaeology in Australia is to some extent defined by legislation. The Maritime Archaeology Act, 1973, for example aims in Western Australia 'to make provision for the preservation on behalf of the community of the remains of ships lost before 1900, and of relics associated . . .' Thus a part of our role as public maritime archaeologists is to save these national treasures—for their intrinsic value, not just for their value in answering researchers' questions. This object-oriented emphasis is a different approach from that of the academic archaeologist and explains the keen interest of the lay public in the fieldwork carried out by maritime archaeologists. Asked about why they have come on a government-organised field expedition as a voluntary assistant, few would answer 'to assist in the formulation of universal rules about man's behaviour', but all would be interested in finding out about history from artefacts.

It would be a blinkered maritime archaeologist who sought information about vessel construction exclusively from shipwrecks on the seabed while ignoring similar vessels still afloat or on shore. Information about ships and boats can come from a variety of different types of sources, each with its advantages and disadvantages. Thus, a shipwreck represents to some extent a moment in time (with modifications up to that moment), whereas a vessel still afloat will have continued to be modified. Several of the State museums have collections of vessels representing aspects of Australian history. Ship museums provide a useful source of new ideas about aspects of vessel construction for maritime archaeologists, given that there is a full appreciation of the way the vessels have been altered during their existence. Museum staff engaged in restoration of such vessels gain parallel insights to the maritime archaeologists conducting excavations on a hull on the seabed. The associations between displayed fragments excavated from shipwrecks and the complete vessels are equally beneficial to the museum-visiting public.

THE TYPES OF UNDERWATER REMAINS AROUND AUSTRALIA

The main areas of archaeological study underwater can be categorised as submerged settlements, harbour works, refuse deposits and shipwrecks. Examples of each of these categories have been found and studied above water in various parts of the world. Particular archaeologists select sites above or below water, depending on the potential of the site for answering questions posed, and upon the ability and predilection of the archaeologist to dive.

Australia has a variety of biophysical environments affecting the creation and condition of underwater cultural sites. It is a continental island with an area of 7.7 million square kilometres (approximately equal to the United States of America), situated roughly between 10° to 45° south and 110° to 155° east, with a 36,700-kilometre coastline (one of the longest coastlines in the world). Our Exclusive Economic Zone, when proclaimed, will cover an area of over 2 million square kilometres of water and be the fourth largest in the world. The shallow seas to the north were dry during the last ice age, but the other coasts have relatively narrow continental shelves. The lack of indentations on these coasts has produced a dearth of suitable harbour sites. The continental shelf is not generally cluttered with outlying islands, although the Great Barrier Reef, girding the north-east coast, and Bass Strait and Tasmania in the south, are major exceptions.

The climate is generally warm because the continent lies in fairly low latitudes. The size of the continent produces temperature extremes inland, but along the coast the ocean modifies temperatures. The southern part of the continent is affected by westerly winds (the roaring forties) while south-easterlies (the trade winds) blow further north. In the summer, because of the overheating of northern Australia, which causes a noticeable drop in atmospheric pressure, the northern edge of the south-east trade-wind belt is truncated, and westerly or north-westerly winds take its place.

During the course of excavation, aspects of the biophysical environments of a number of Australian underwater sites have been recorded. But no attempt has yet been made to classify the body of known Australian underwater sites in this manner. If this were done it would be possible to predict the condition of many of the as-yet unfound or unexplored sites. Keith Muckelroy constructed a table, relating to condition of preservation, of five classes of wreck sites in British waters. On one of Muckelroy's first-class sites there are extensive structural remains, many organic remains, many other objects, and coherent distributions. The Swedish warship *Wasa* (lost at Stockholm in 1628) and the British warship *Mary Rose* (lost at Portsmouth in 1545) would be examples of this class. Similarly, the British Navy's HMS *Pandora*, lost off Queensland in 1791, fits comfortably into this class. The Dutch East Indiaman *Batavia*, wrecked off Western Australia in 1629, would be a second-class site, even though it is currently considered to be Western Australia's most important wreck site archaeologically, and the Dutch East Indiaman *Vergulde Draeck*, wrecked off Western Australian in 1656, would be considered a third-class site because of its poor preservation conditions.

Muckelroy's table was based on topography (% of bottom sedimentary deposit), deposit (range of sediments), slope (average over whole site), sea horizon (sector of open water for 10+ km), and fetch (length of open water where wind or waves might be generated).[8] On his first-class site one expects 100% of bottom sedimentary deposit, a range of sediments between gravel and silt, a minimal slope over the whole site, an open-water sector of less than 90° and a fetch of less than 250 km. For Australian sites it would be useful to add to the table consideration of such influences as water temperature, water depth and tide velocity. The table was constructed for shipwreck sites, but to a lesser degree it is also applicable to other underwater sites.

Australia's underwater sites could be classified in other ways also. Predictive models (devices to help explain theories) could be developed, for instance, to illustrate the observed relationship between variables influencing shipbuilding—and applied to groups of India-built shipwrecks, or slave-ship wrecks, or pearling lugger wrecks.

A. Settlements

Because of geological upheavals, erosion, and water level changes over time, cities or smaller settlement sites in many parts of the world have been inundated with water. This may have occurred gradually, inducing succeeding generations of inhabitants to move their houses and belongings further up the bank, or cataclysmically, resulting in great loss of life and preventing any salvage of possessions taking place. Port Royal, a thriving city in Jamaica, suddenly disappeared beneath the waves during an earthquake on 7 June 1692. In view of the element of surprise the disaster may be likened to that which occurred at the Italian town of Pompeii, buried by an eruption of Mount Vesuvius in AD 79 or to the landslide that occurred in Bogota, Columbia, on 15 November 1985. On the days of these three catastrophies time suddenly stopped for some of the inhabitants, whose lives ended as they were drowned, petrified or buried in mud. Abandoned disaster sites are exciting material for archaeologists, providing the opportunity to study life as it was on that day, without having to consider the modifications and 'contamination' of succeeding generations of occupants.

During the brief period of European habitation of the Australian continent there have been no major floodings of settlements by water, except on a seasonal basis or as a result of deliberate planning for dam projects. But while Australia was the last continent to be inhabited by Europeans, it has been colonised by Aborigines for over 40 millennia.[9] The first Aborigines came here from south-east Asia during the last glacial epoch, when lower sea levels extended the land mass of Australia to result in land bridges with Tasmania and New Guinea.

Australia was not joined to Asia. To reach Australia people from Asia had to cross several water barriers, and the minimum sea distance was never less than about 100 kilometres. Only groups with some familiarity with the sea could possibly enter Australia. According to prehistorian Rhys Jones 'the key to Australian prehistory lies in our understanding of the ability of Aborigines and their far distant ancestors to cross water'.[10] He also notes that this is the oldest evidence anywhere in the world for such maritime journeys. So the first people to enter Australia were tropical Asians with a technology which embraced seafaring and woodworking, including boat construction.

It would have been easiest to cross the water barriers when the sea levels were lowest, and this has influenced the prehistorians' theories. It has been argued that the colonisation pattern in Australia was a migration along the shelf.[11] The earliest migrants would have come from the Indonesian island chain and had cultural adaptations to island coastal and littoral environments. So rather than move into the arid interior of Australia, population expansions would have occurred along the coast.

When the ice receded in northern Asia and in Antarctica the climate of Australia changed and at the same time the level of the ocean rose until Tasmania, Kangaroo Island, Australia, New Guinea and the islands of Indonesia and the Malay Archipelago were cut off from each other by expanses of sea.

The earliest Aborigines would have left implements as they crossed, occupied and retreated from ground now covered by the waters of Bass Strait, Torres Strait and the Timor Sea. Artefacts found in stratified archaeological deposits in these areas would give us a better idea of the dates when the Aborigines first crossed to what are now the Australian continent and the island of Tasmania. They would also give us a better idea of how coastal-dwelling Aborigines lived. Today's non-diving archaeologists can only study on land those traces of occupation which were 100 metres or more above sea level, and often many kilometres inland, at the time of the key crossings. On Hunter Island in Bass Strait excavation has been carried out on an Aboriginal site which dates to the time when there was a land bridge connecting Tasmania with the mainland, and Hunter Island was Hunter Hill.[12] But traditional archaeological techniques are unable to help in the search for evidence of occupation of areas now inundated, because the Aborigines were generally hunter-gatherers who led a more or less nomadic life, and burdened themselves with few personal effects. Thus the implements will be scattered widely, few in number and limited in variety. Archaeologists may well have to rely on the trawl nets of fishermen and oceanographic expeditions for the first clues in the search for early Aboriginal artefacts in these waters.

The Institute of Oceanographic Sciences in Britain recently commenced an ambitious five-year project to find submerged sites off Australia, the ideal being the location of the remains of the first human migrants who crossed the northern continental shelf of Australia from Indonesia. The first phase, a diving survey in the Timor Sea at the Cootamundra Shoals, 240 kilometres north west of Darwin, was conducted in 1983 and found potentially habitable terrain.[13]

A recent hypothesis regarding the colonisation of Australia is that rising sea levels in Indonesia could have provided impetus for migration.[14] Such circumstances could

The Australian continent.

well have induced people who did not normally look to the sea for their food to take to the water. These people, on reaching Australia, might have reverted to their earlier way of life, away from the water. Because of the great distance in time, such questions cannot easily be resolved.

B. Harbour Works

The study of harbour installations has revealed a good deal about the development of sea power and trade of early Mediterranean countries. The record preserved in many of these stone blocks on the seabed, such as for example the completely unknown and perfectly preserved Phoenician harbour found at Atlit in Israel in 1964, is not rendered obsolete by archival information. In Australia the larger ports have continued to develop since their construction in the eighteenth and nineteenth centuries. In some cases archaeological techniques such as ground surveys and the study of aerial photographs can be expected to throw further light on the history of these still surviving structures.[15] The massive redevelopment in progress at Darling Harbour, Sydney, should provide the opportunity for examination of a variety of early port features, both above and below water. A study of the smaller outports could also be helpful. No port records have survived of many of the scattered jetties and mooring areas where sailing vessels loaded the produce of the land or sea for export. This applies particularly to single industry settlements, such as the Norwegian Bay whaling station on the north-west coast of Western Australia, brought into prominence by boom conditions and then abandoned as the economic cycle turned. Navigation aids including lighthouses, such as the one on Raine Island, Queensland, are another potentially useful archaeological resource. Automatic beacons have all but replaced the traditional lighthouses and the only intact examples may soon be those in museums. One such lighthouse, brought to Australia in 1869, was operating at South Neptune Island until 1984, and has now been transported to the South Australian Maritime Museum at Port Adelaide.

C. Refuse Deposits

Associated with jetty and mooring areas are the inevitable collections of jetsam. The archaeological information obtained from underwater refuse sites in Australian waters has been limited, resting on individual objects rather than on assemblages of artefacts. On some occasions these sites can be expected to make a modest contribution to the record of Australian history. The numerous coins, items of jewellery and other personal belongings raised by divers from the seabed off the Glenelg jetty in South Australia have indicated clearly that the inward passengers from the passing Peninsular and Oriental (P&O) Company steamers were more than a little inconvenienced by having to transfer from the moored steamer to a boat for the run into the jetty, and then having to clamber from the rocking boat up the jetty with their baggage. But this confirms what can perhaps alternatively be learnt from the available documentary sources.

Examination of the seabed around the remains of the Lockeville jetty on the south-west coast of Australia, an area used extensively by sailing ships carrying away timber in the late nineteenth century, reveals an assortment of the general maintenance tools employed on vessels of that period. Similarly it might be expected that the seabed adjacent to the early nineteenth century convict establishment on Sarah's Island off Tasmania would yield a range of material associated with the convict system.

D. Shipwrecks

The shipwrecks in Australian waters have well-recognised potential for archaeologists. Australia has a particularly long coastline and a small population, so people placed a great reliance upon shipping for trade and communications in the early days of European settlement. Thousands of vessels were wrecked under a variety of circumstances, providing maritime archaeologists of today with a wealth of source material. It is now known that in the Mediterranean area seafarers (and therefore wrecked watercraft) of 10,000 to 12,000 years ago preceded farmers and shepherds.[16] Man reached Australia in watercraft at least three times as far back in time. Aborigines of the Pleistocene period must have lost some of their substantial sea-going craft. This is the earliest evidence, indirect though it is, anywhere in the world for wrecked watercraft. None of the watercraft types now known from ethno-historical and ethnographic records to have been used by traditional Aboriginal people seems likely to have been capable of extensive sea crossings, and it seems improbable that remains of craft used by the earliest Pleistocene colonists will be found.[17] They were too perishable to have survived. The earliest known wreck site is European, and occurred in 1622. In order to understand the reasons for that and following shipping casualties it is necessary to trace the movements of earlier European traders who ventured into the Indian Ocean.

Chapter 2   REFERENCES

1. Bass, 1966, p. 15.
2. Bass, 1980, p. 143.
3. Muckelroy, 1978.
4. Muckelroy, 1978, p. 4.
5. Gould, 1983, p. 7.
6. Thomas, 1974, p. 2.
7. This argument is taken from Seasholes, 1984, pp. 12-13.
8. Muckelroy, 1978, p. 164.
9. Shawcross, 1975, Mulvaney, 1979, Pearce and Barbetti, 1981, p. 173.
10. Jones and Meehan, 1977, p. 16.
11. Bowdler, 1977, pp. 205-246.
12. Bowdler, 1976, pp. 29-33.
13. Flemming, 1985, pp. 19-25; Flemming, 1984. See also *Cootamundra News*, 1982-3.
14. Fairbridge, 1983, p. 615.
15. Department of Home Affairs, 1980.
16. Bass, 1980, p. 137.
17. Flood, 1983, p. 36.

3 Ships in Australian Waters Before European Settlement

The first steps towards the discovery of Australia by a European nation were taken by the Portuguese. In 1497 Vasco da Gama, sailing under the patronage of King Manuel the Fortunate, rounded the Cape of Good Hope and continued eastward as far as India, reaching the great spice port of Calicut in 1498. Having found the means of circumventing Arab control of the overland trade in spices to Europe, the Portuguese quickly consolidated their position, capturing the island of Goa in 1510 and Malacca in 1511. They then had the necessary bases from which to organise their own trade route through the Indian Ocean, and they reached a principal source of the spices, the Moluccas, in 1511-12.

Portugal's trading ventures brought her ships to regions quite close to Australia. Historians are divided over the likelihood of a sixteenth century Portuguese discovery of the fifth continent. Mercator's 'Terra Australis' is perhaps a derivation of Ptolemy's armchair concept of a southland to counterbalance the great land masses of the northern hemisphere. But the French maps of the Dieppe school, which appeared in the mid-sixteenth century, appear to have come from Portuguese sources and some historians have taken the view that the Java la Grande which figures in these maps is in fact the Australian continent. No substantive evidence for a sixteenth century Portuguese discovery has come to light.[1]

The balance of power in the East was to shift after the second half of the sixteenth century as Portuguese areas of control of trade were threatened firstly by the Spanish, and later by the Dutch and English.

Many a Dutchman had sailed to the Indies on Portuguese ships. Cornelis de Houtman initiated direct Dutch involvement in the sea route to the Indies in 1595, when he sailed a fleet of four ships to Bantam. In 1602 the Dutch trading companies united to form the Verenigde Oost-Indische Compagnie (V.O.C.), which obtained the exclusive rights to send fleets round the Cape of Good Hope and through the Straits of Magellan.

This company was formed with the intention of trading with the spice islands and ultimately driving all other European powers from the trade with the East Indies. The Dutch quickly asserted their rule of the sea; they enforced their nutmeg monopoly on the Banda Islands in 1602, founded the first trade factory in Bantam in 1603, established a protectorate over Amboina in 1605 and in 1609 began a system of con-

The Cape of Good Hope, a welcome pause on the outward journey.

solidation through direction by Governors General in the East. Batavia (present-day Jakarta) became the centre of Dutch trading interests which already extended to the Moluccas in the east, China and Japan in the north and the Coromandel Coast and Surat in the west. So thoughts of expansion to the south were imminent.

In 1605 Jan Willemsz Verschoor, in charge of Dutch trade in Bantam, sponsored a scheme 'to discover the great land Nova Guinea and other unknown east and south lands'.[2] The undertaking was entrusted to Captain Willem Jansz, who in the pinnace *Duyfken* sailed from the Banda Islands to the south coast of New Guinea then south east to the Australian coast, encountering the west side of Cape York Peninsular at the Pennefather River (11°45′S) early in 1606. Jansz sailed some 320 kilometres along the Australian coast without discovering Torres Strait. At the mouth of the Batavia River he had an unpleasant encounter with the Aborigines:

> ... in sending their men on shoare to intreate of Trade, There was nine of them killed by the Heathens, which are man-eaters; so they were constrained to returne, finding no good to be done there.[3]

Jansz had made the first known landing by a European on the Australian continent, but his discovery was of little immediate consequence: there seemed to be no prospect of trade there, and the newly discovered lands did not lie in the path of any of the then established sea routes.

The numerous landfalls which were to occur on the west coast of Australia may be attributed to the sailing instructions for the route between the Netherlands and the East Indies, issued in 1617. These became an important landmark in the history of navigation.

The Dutch pioneers had at first used charts obtained from Portuguese sources, and followed Portuguese sea-routes to the East. After rounding the Cape of Good Hope they usually sailed a north-easterly course up the east coast of Africa through the Mozambique Channel and then across to India. Later as rivalries intensified they avoided the Portuguese stronghold in Mozambique by sailing up the east coast of Madagascar as far as they could, often touching Mauritius and then making directly for the Sunda Straits.

The route offered many advantages for ships returning from India, but brought hardships for vessels sailing on the outward journey. Outward bound vessels had to sail laboriously against the prevailing south-easterly trade winds, and after that were frequently faced with long periods becalmed in the doldrums, when the scorching sun would melt the pitch in the planking of the deck, the cargo would begin to decay, the drinking water become foul and the crew be decimated by scurvy. Navigational hazards in the form of numerous shoals, islands and reefs in the central Indian Ocean added to the worries of the Dutch commanders.

Hendrick Brouwer drew the attention of the Dutch company to the potential value of the westerly winds which blow all the way round the globe between latitudes 35° and 40° south. In December 1610 he was sent out to the East Indies. After rounding the Cape of Good Hope he sailed south to beyond latitude 36° and then set a course due east, turning north only when he believed himself to be in the same longitude as the Sunda Strait. Finding a favourable wind he soon reached the Strait and Batavia.

The voyage from Holland to Java took Brouwer just under six months, in contrast to the year or so usually needed to complete the journey by the traditional route. Brouwer succeeded in harnessing the strength of the Roaring Forties, and by staying in cooler latitudes provisions were kept fresh and the crews healthier. Other captains investigated the route and reported favourably, and as a result the Company instructed commanders to lay their course east from Table Bay at between latitudes 35° and 40° south, depending on where they found the best winds. After the westerly winds were found, ships were to sail eastwards for at least 1,000 Dutch miles before turning north, since there were no navigational hazards on the Javanese coast, whereas vessels altering course before covering 1,000 miles ran the risk of being drawn off course to the shores of Sumatra, where vessels might be becalmed for long periods.

For several reasons it was difficult to measure accurately the 1,000 miles to be sailed eastwards. Navigation was practised on the basis of reckoning. Estimation of a ship's position was made by three procedures: observation of the sun, using an astrolabe, giving the latitude; calculation from the log and sand glass, giving the distance covered; and examination of the compass, giving the course steered. The navigator would depart from a point of known latitude and longitude on a predetermined compass course. He would prick off on the chart his new position at noon the following day, on the basis of the course sailed and the estimated distance, which was corrected in terms of latitude by observation of the sun. Longitude had to be estimated by dead reckoning alone. Dead reckoning is the calculation of a ship's position from the distance run by the log and the courses steered by the compass, with corrections for current, leeway, etc., but without astronomical observations. So although the reckoning could be corrected daily in terms of latitude, the error in longitude accumulated as the voyage progressed.

Confusion resulted from the fact that several lengths were attached to the term 'mile' around the time the instructions were given for the new route, leading some captains to sail further than others. Another uncertainty resulted from the use at the time of both plane charts and Mercator charts. With these complications and uncertainties it was inevitable that soon after ships began using the new route some would come across the coast of Australia.

The first sighting of the west coast of the Southland was made in the *Eendracht*, which left Holland for Java in January 1616 under the command of Dirck Hartog. Leaving Table Bay in August he began following the new southern route as recommended by Brouwer, but delayed too long in turning north for Java and came upon the coast of Australia, in the vicinity of what is now known as Shark Bay. Hartog nailed a plate to a post to mark the occasion, then sailed north to Bantam.

Thereafter, Dutch sightings of the coast followed in quick succession. In May 1618 Haevik Hillegom, captain of the *Zeewolf*, saw land at 21° 15' south, an area near the North West Cape. The *Mauritius*, under the command of Lenaert Jacobszoon, came unexpectedly upon the same stretch of coastline 3 months later. By coincidence the supercargo of the *Mauritius* was Willem Jansz, who as captain of the *Duyfken* had been the first 'discoverer' of Australia.[4]

The *Dordrecht* and the *Amsterdam*, part of a squadron commanded by Frederik de Houtman on board the *Dordrecht*, came suddenly upon the Southland in July 1619. Houtman first saw the coast a little south of the Swan River and then sailed northwards past the group of islands now known as the Houtman Abrolhos. The *Leeuwin* sighted the coast even further south in March 1622, discovering Cape Leeuwin, the south-western-most point of Australia.

Ultimately the use of the southern route was to lead to the loss upon Australian shores of at least four Dutch and one English East Indiamen during the seventeenth and eighteenth centuries.

The first victim was to be an English vessel, the *Trial*. The English had entered the trade with the East at about the same time as the Dutch. In 1591-3 the *Edward Bonaventure* commanded by Captain James Lancaster was the first English ship to pass the Cape of Good Hope and cross the Indian Ocean to the Malay Peninsula. The English East India Company (E.E.I.C.) received its charter in 1600 and sent out its first fleet under the command of Lancaster in 1601-2.

Brouwer's route could not be kept secret from the English for long. In 1620 Captain Humphrey Fitzherbert was placed in command of a fleet of three E.E.I.C. ships bound for Bantam. Arriving at the Cape of Good Hope in the *Royal Exchange*, Fitzherbert became friendly with the Dutch commander of the V.O.C. ship *Schiedam*, who advised him of Brouwer's route. Fitzherbert's voyage across the Indian Ocean to Java lasted a mere 7 weeks.

The following year the English company sent out a number of vessels to the East, including the *Trial*. Little is known of the vessel itself. She was a ship of 500 tons, with a crew of 143 men, and was said to carry her sails well. The Company decided to send to the Indies in the *Trial* small items such as sheathing nails, hunthorns, cartridges and sheet lead. She also carried spangles for the King of Siam and 500 silver reals (a small amount of coinage for an outward bound East Indiaman) intended for general provisioning of the English establishment at Batavia, and for the payment of the English share of fort charges at Pulicat near Madras.

The *Trial* sailed boldly out of Plymouth with a full cargo on 4th September 1621, under the command of Captain John Brookes. She arrived safely at the Cape of Good Hope but it was here that Brookes showed himself to be uneasy about the prospect of a long journey in high latitudes across the unknown southern Indian Ocean. He had a copy of Fitzherbert's log but that voyage, only a year previous, had been the very first along the new route by an English vessel, and neither Brookes nor any of his officers had ever been to Batavia before. He tried to persuade the first mate of the *Charles*, a returning East Indiaman, to accompany him to the Indies, but being unsuccessful he left the shelter of the Cape on the last leg of his ill-fated voyage on 19th March, 1622.

Following Fitzherbert's journal Brookes sailed south to latitude 39° and then east across the great expanse of ocean towards Australia. But Brookes let the winds take him further east than Fitzherbert had deemed prudent, and on 1st May he sighted the Australian mainland, in latitude 22° south, the region of North West Cape where in earlier years several Dutchmen had paused to wonder at this great unknown land.

North-easterly winds at first prevented Brookes from setting a course for Java, but on 25th May he was again heading north-east, past Barrow Island and the Monte Bello Islands towards the uncharted Trial Rocks. At 11 o'clock that night, during fair weather and smooth water, the *Trial* crashed unexpectedly onto the reef. Brookes ran to the poop, and sounding, found only 6 metres. The steersman tried tacking to the westward, but it was of no use and the ship rapidly filled with water. The crew could see neither land nor reef, but the ship was stuck fast in 4 metres. The wind freshened and it became clear that the ship could not be backed off. Brookes had his skiff launched to sound around the ship.

He left the *Trial* in the skiff at 4 a.m. and half an hour later the fore part of the ship fell apart. His hurried departure was witnessed with contempt by the first mate Thomas Bright, who claimed that Brookes:

> like a Judas running into [the] great Cabin lowered himself privately into the skiff with only nine men and his boy, stood for the Straights of Sundaye that instant without care.[5]

With great difficulty Bright launched the longboat and 36 men tumbled in before he was able to cast off from the ship. He remained a quarter of a mile from the *Trial* until daylight, reluctant to desert his fellows but determined not to let the boat be overturned by the desperate scramblings of those 97 men who remained to die on the wreck. With the sun came rising seas so the longboat crew set out for the Monte Bello Islands, lying low on the horizon. Here the men searched unsuccessfully for water before turning northward towards Batavia.

Brookes's skiff had on board one keg of water, two cases of bottled wine and a little bread, while the longboat under the command of Thomas Bright had six kegs of water, a little wine and some bread. Brookes reached the eastern end of Java on 8th June, and arrived at Batavia on 25th June, a month after the *Trial* struck. Here he claimed that he had ordered the Company's money, gold spangles and ship's papers, together with his own money, to be placed in one of the boats, but that the whole lot had been lost overboard. Three days later the unexpected arrival of Bright's boat, which had taken an independent course, brought challenges to the captain's veracity, for it was alleged that Brookes had stolen goods from the Company and been negligent in his handling of the ship. Bright claimed that Brookes had spent the first two hours after the *Trial* struck in transferring to his own chest the Company's money, spangles and letters, and in deliberately heaving overboard those documents which might incriminate him.

It is now apparent that Brookes falsified his journal to give the impression that he had not taken his vessel further east than Fitzherbert had gone in 1620.[6] Brookes's falsification placed the site of the wreck many miles west of its true position and as a result the Trial Rocks were to remain unidentified for over 300 years. In 1934 it was finally established that an isolated reef then known as Ritchies Reef, lying about 130 miles north east of North West Cape, was in fact the rock on which the *Trial* had been lost. But the E.E.I.C. had no way of knowing this and were duped by Brookes's elaborate lie: his story was entirely false, but completely consistent. So Brookes was not found guilty of negligence, and he gained a fresh command soon afterwards.

The immediate significance of the wreck of the *Trial* related to the Company's money, either stolen or lost by Brookes. The tenuous hold of the English upon their small outposts in the East was dependent upon the regular supply of money to meet their local commitments. Thus Brookes, by his inept handling of the *Trial*, played some small part in the events leading to the demise of the English company in the spice islands. By 1623 the E.E.I.C., lacking the government support enjoyed by the Dutch company, had been ousted from the Archipelago. It had found a new and profitable field of enterprise, at the expense of the Portuguese, in India, to which Captain Thomas Best led a fleet in 1612 and where commercial relations were established by Sir Thomas Roe's mission to the Great Mogul in 1615–18.

The Dutch, on hearing of the loss of the *Trial*, took steps to learn more about this long, barren coast which sometimes blocked their passage. By 1627 they could piece together the Australian landfalls to make the almost continuous coastline shown in the Hessel Gerritz chart, stretching from De Witt's land in the north west to Pieter Nuyt's land and the Isles of St Francis and St Peter in the Australian Bight.

But these cartographical advances were not to prevent the disaster which befell the V.O.C. ship *Batavia* in 1629. The *Batavia* sailed from Amsterdam on her maiden voyage to the East Indies in October 1628, commencing one of the most tragic chapters in Dutch maritime history.[7] She was the flagship of a fleet of three vessels when she left Holland. Francisco Pelsaert, president of the fleet, was on board. The vessel carried the usual trade goods and supplies, such as cloth, lead and cochineal, wines and cheeses, and 12 chests of silver coins worth 250,000 guilders for the purchase of goods in the Indies. The *Batavia* also carried a quantity of jewellery. This included a casket of jewels valued at 58,000 guilders, some wrought silver work, and a fabulous gem known as the 'great cameo', made by the painter Rubens and intended for sale to the Emperor Jahangir, son of Akbar the Great Mogul.

During the voyage the supercargo, Jeronimus Cornelisz, conspired with the pilot and other officers to seize the ship and engage in piracy. The ship was close to mutiny when on the night of 4th June 1629 she crashed onto the coral reef surrounding the Wallabi Group of the Houtman Abrolhos.

The skipper, Ariaen Jacobsz, had seen white froth on the water some distance from the reef but members of the night watch assured him it was merely the reflection of the moon on the water. Upon sounding they found 6 metres of water aft and much less forward, so the cannon were made ready to be thrown overboard to lighten the ship, and the mainmast was cut away. In the morning a boat examined the nearby islands and most of the crew were sent on shore to pacify the women and children, and the sick.

More than 250 of the people on board managed to reach the islands, but 40 were drowned in the attempt. Little water was saved from the ship so after an unsuccessful search of the other islands and the mainland by boat, Pelsaert and Jacobsz, with 46 crew, decided to leave the others and try to reach Batavia for help. After a difficult journey northwards they sighted Java on 27th June and made their way westward to Batavia.

Meanwhile back at the islands the supercargo, Jeronimus Cornelisz, had remained upon the wreck for 10 days after the vessel struck, having found no means of reaching the shore. He then spent 2 days clinging to the small bowsprit mast, which floated eventually on to one of the islands. In the absence of Pelsaert he became the leader of the group, and being perhaps more than a little disillusioned with his absent commander he deemed this to be a suitable occasion for putting his original plan into operation. When the rescue ship arrived he would surprise the commander and seize the vessel, after which he would cruise as a pirate in the East. In order to achieve this villainous end it was necessary to get rid of those of the crew who were not loyal to him.

The crew of the *Batavia* were divided between three islands: Traitor's Island, so named after the departure of Pelsaert; the island named Batavia's Graveyard, where Cornelisz and the greatest number of men and women were camped; and another island where a group of men under the leadership of Wiebbe Hayes had been sent under pretext to search for water. Hayes and his group of 45 soldiers found an abundant water supply after searching for several weeks, but when Hayes made the arranged signal to confirm with Cornelisz that he had found water, he was ignored. Cornelisz meantime, in an orgy of rape and murder, had done away with 125 of those on the other two islands who were not of his party of conspirators. One man managed to escape and joined Hayes on his island to inform him of the massacre that had taken place.

Cornelisz mounted two assaults on Hayes's island but was repulsed. A new tactic was then tried, whereby a treaty of peace was proposed: Hayes would be unmolested, and receive some clothing, provided he deliver up his small boat to Cornelisz. During the negotiations Cornelisz, too cunning for his own good, wrote to some of Hayes's men, trying to corrupt them with bribes. The letter was shown to Wiebbe Hayes, so the next day when Cornelisz arrived to continue the negotiations Hayes's men killed Cornelisz's several bodyguards and made Cornelisz himself a prisoner.

Soon afterwards Pelsaert arrived from Batavia in a rescue ship the *Sardam*, and cast anchor. He proceeded towards one of the islands in a boat but was intercepted by Wiebbe Hayes in another boat, who warned him to return to his ship to avoid being surprised by the remaining conspirators. Pelsaert then saw Cornelisz's two boats coming towards him, and scarcely had time to re-embark upon the *Sardam* before Cornelisz's men came alongside, armed and wearing costumes embroidered with gold and silver. Forewarned, Pelsaert threatened to sink their boat unless they submitted, and they were immediately placed in irons.

On 18th September the remaining mutineers were rounded up and a quantity of jewellery which had been in their hands was recovered. The wreck was visited and found to have been ripped apart by the savage seas. The keel lay on a sandbank in one direction, while the foremast had come to rest on a rock in the opposite direction. The steward told Pelsaert that he had gone fishing on a fine day and seen one of the silver chests on the wreck. Indian divers recovered 11 of the chests leaving one, which they found impossible to move, marked with an anchor and a cannon. Pelsaert, anxious to please the V.O.C., scoured the islands for every salvageable vestige of the Company's goods, collecting even barrel hoops. The mutineers were tried, sentenced and, in some cases, executed on the islands before the *Sardam* left for Batavia. Two of the conspirators, Wouter Loos and Jan Pelgrom, were sentenced to be marooned on the mainland, and accordingly they were put ashore at an inlet at 27° 51', probably the mouth of the Murchison River.

The *Vergulde Draeck*, popularly known as the *Gilt Dragon*, set sail from Texel on 4th October 1655 on her second and final voyage to Batavia for the V.O.C. She sailed under the command of Pieter Albertsz with a crew of about 193, and a cargo of trade goods worth 106,400 guilders as well as eight chests of silver coins worth 78,600 guilders.[8]

On the way out to the Cape of Good Hope two men were lost overboard and a third presumably died, as the numbers were reduced to 190. Following a brief refresher visit at the Cape she sailed east, following Brouwer's route, but struck a reef off the coast of Western Australia in 31° 16' south latitude a little before first light on the morning of 28th April 1656. The vessel immediately burst open and sank, leaving only her stern projecting above the water. Two boats were launched but only 75 men managed to reach the shore, leaving the rest to drown on the reef or in their bunks below decks along with the rats. All that could be saved from the ship were a few provisions. The understeersman and six men were dispatched to Batavia to seek assistance, while the master, Albertsz, mindful of the fate of the leaderless *Batavia* wreck survivors, remained on the shore with the rest of the men.

The boat arrived at Batavia on 7th June and the crew reported that as they had sailed away they saw the other survivors trying to refloat their boat, which had capsized on landing. A number of vessels were sent to search for the remaining survivors and try to salvage the Company's treasure, but without success. In 1658 one party of searchers, in a boat from the *Waeckende Boey*, were themselves marooned on the coast when their ship sailed off and left them. Abraham Leeman, in charge of the boat-load of 14 men, steered north for Batavia, arriving there with three surviving members of the boat crew on 23rd September, 185 days after having been marooned on the Australian coast.[9]

The fate of the *Vergulde Draeck* survivors left on the beach remains a mystery. It seems likely that some of them succeeded in launching their boat, as the later search parties did not report seeing it in its original position on the beach. If the boat was launched then it must have subsequently capsized between there and Batavia. Others must have stayed on the Australian coast because there was not room for everyone on the boat. In 1931 a boy, A. Edwards, found about 40 silver coins and fittings from a small chest in the sand hills on the coast adjacent to the wreck. As the dates on the coins ranged between 1619 and 1655 it is probable that they were brought ashore from the wreck by the survivors.[10] It is highly unlikely that the Company's servants would have abandoned this specie, so it would seem that those left on the coast lingered and died in the area adjacent to the wreck.

Following the wreck of the *Vergulde Draeck*, 56 years elapsed before the next V.O.C. ship was lost on the Australian coast. The *Zuytdorp* sailed from Flushing on 27th July 1711 on her third voyage from Holland to Batavia. No cargo list has survived, but other records indicate that the vessel carried about 250,000 guilders in cash, mainly silver but perhaps some gold.[11] The *Zuytdorp* was under the command of Marinus Wysvliet, an experienced skipper of the Company. During a slow 7-month voyage to the Cape of Good Hope 112 of the crew of 286 died, and 22 became ill. On 22nd April 1712 the *Zuytdorp* departed from the Cape, bound for Batavia in company with the *Kockenge*. The *Kockenge* arrived at Batavia on 4th July, but the *Zuytdorp* was never seen again. No searches were made because there was nothing to indicate where the ship was lost.

The last of the V.O.C. ships known to have been wrecked on the Australian coast was the *Zeewijk*. This vessel, commanded by Jan Steyns and manned by a crew of 212 men, departed Holland in November 1726, bound for Batavia with a rich cargo including silver coins to the value of 315,836 guilders carried in 10 chests.

The *Zeewijk* safely reached the Cape in March 1726 and soon departed for the voyage across the Indian Ocean. But Steyns, an unseasoned skipper, was tempted to catch a glimpse of the unknown Southland. Disobeying the Company's strict instructions and ignoring the protests of the steersman he kept his course to the east well after the point where he should have turned northwards.[12]

A model of the VOC ship *Zuytdorp*, by Dr J. de Heer. (Photo: Pat Baker)

Two hours after sunset on 9th June the ship suddenly piled up on the reef skirting the western side of the Pelsaert group of Houtman Abrolhos. The sailor on lookout had been watching the breaking surf for half an hour, but thought that it was a reflection from the sky.

The stoutly built vessel held together on the reef for some months, despite the punishing ocean swells which for a week prevented the longboat from even being launched. A camp was established on a nearby island (later to be named Gun Island) and the crew began salvaging valuable and useful items from the wreck. The longboat was put into seagoing order and 11 of the best seamen under the command of the first officer, Pieter Langeweg, set out for Batavia to seek help. Nothing more was heard of them.

A small sailing vessel named *Sloepie* was constructed out of wreckage salvaged from the *Zeewijk*, and in March 1728, nine months after their ship had been wrecked, the survivors left for Batavia. Eighty-two men arrived at Sunda Strait in this vessel on 21st April.

The hapless skipper, Jan Steyns, was prosecuted for his carelessness in sailing too close to the Southland, and for endeavouring to falsify the journals to hide his mistake. He was convicted on all counts and it was decreed that he should have his goods and money confiscated, be deposed of office and banished for life from all the places under the Company's administration, and that he be exhibited tied to a pole at the place where criminals were publicly executed, wearing a board around his neck on which the word 'falsifyer' was written.

Other Dutch ships are known to have been lost between Cape Town and Batavia. In 1619 the *Wapen van Amsterdam* was wrecked on the south coast of New Guinea, and some of the crew were killed by the natives. At that time the Dutch were not aware that Torres Strait separated New Guinea from Cape York Peninsula, so it cannot be definitely stated that the vessel was not wrecked in Australian waters, although the north coast of Australia hardly matches the Dutch description as 'the south coast of Nova Guinea'.[13]

Another loss was the V.O.C. ship *Ridderschap van Holland*, which sailed from Flushing in July 1693 bound for Batavia via the Cape of Good Hope. She left the Cape on 5th February 1694 but never reached Batavia. In November 1695 the directors of the company in Amsterdam resolved to send out a search party. Willem de Vlamingh was placed in command of a flotilla consisting of the frigate *Geelvinck*, the hooker *Nijptangh* and the galliot *Weseltje*, and he searched along a route including St Paul and Amsterdam Islands and the coast of Western Australia between Rottnest Island and North West Cape without success.

Later reports, however, indicated that the *Ridderschap van Holland* was wrecked near Madagascar. After it had sailed from Table Bay a mast had broken and the skipper Dirck de Lange steered the vessel on a course for the nearest land, Madagascar, where she was lost with all hands. The crew were killed by pirates, probably upon the orders of the Jamaican mulatto, Abraham Samuells, who by 1695 had risen to be the acknowledged leader of all the pirates at Fort Dauphin on the Madagascar coast.[14]

The V.O.C. lost many ships during the 1720s including several on the African coast and three; the *Fortuyn*, the *Aagtekerke* and the *Zeewijk*, between the Cape of Good Hope and Batavia.[15]

It has been conjectured that the *Fortuyn* was wrecked on the Australian coast, but this view is not supported by the contemporary Dutch documents. After a fast and healthy passage outward from Holland the *Fortuyn* anchored in Table Bay on 2nd January 1724, reporting one man dead and three sick. Having taken on fresh provisions she sailed from the Cape on 17th January, bound for Batavia, but was wrecked *en route*. The ship *Graveland*, which left the Cape a fortnight after the *Fortuyn*, came across the floating remnants of a Dutch ship in 13° 20' south latitude. The derelict was sighted on the 6th and 7th April, when the *Graveland* must have been in the vicinity of Cocos Island. An expedition was fitted out in Batavia to look for survivors on Cocos, but did not land on the Island and found nothing.

The *Aagtekerke*, which went missing 2 years after the *Fortuyn*, is more interesting, from an Australian point of view, because it has never been established with any certainty that she was not wrecked on the Australian coast. After obtaining crew replacements and provisions at the Cape, the outward bound *Aagtekerke* put to sea on 23rd January 1726, never to be seen again. By October of that year the Company considered the vessel lost, but it was not until after the wreck of the *Zeewijk* that any information regarding the locality of the wreck was obtained.

The *Zeewijk* survivors, who lived on the Abrolhos for about 10 months, explored a number of the islands in search of water. They reported having seen:

> . . . some signs of a Dutch ship, probably wrecked against the abovementioned reef, which might have been the *Fortuyn* or *Aagterkerke*, whose crew may have died or perished at sea on their way thither.[16]

The seamen spoke of finding a figurehead on Pelsaert Island, and later mentioned a wreck. Yet their evidence remains ambiguous. It is likely that they assumed they had located a wreck because they had found the figurehead, which might well have drifted for hundreds of miles before being wrecked on the Abrolhos. If the *Aagtekerke* does lie on the Abrolhos then the likelihood is that it will be discovered by skin-divers in the near future.

It is possible that other Dutch ships were wrecked on the Australian coast during the seventeenth and eighteenth centuries. Certainly a number which left the Cape of Good Hope remain unaccounted for, but to date no positive evidence has been found of such vessels having been wrecked here.

Maintaining a presence in the East was a costly exercise for the Dutch people. Of the one million people who left Holland for Asia on board East Indiamen during the time of the V.O.C. no more than one out of every three ever returned alive. Yet few settled down in Asia.[17]

Legends abound of Chinese, Portuguese and Spanish wrecks of the fifteenth and sixteenth centuries which are supposed to lie in Australian waters. The most frequently cited is that of an unidentified vessel known as the 'Mahogany Ship' on the beach near Warrnambool in Victoria. This wreck, described in local folklore as a vessel of antique construction, believed to be a galleon, is said to pose a problem of the first magnitude in the controversial history of the discovery of Australia by European navigators. Wreckage was apparently first found there by two shipwrecked sailors in 1836. Subsequent witnesses during the nineteenth century (prior to the wreck's disappearance under the sands) agreed that the wreck lay 300 to 400 metres above high water mark. As the sand dunes are creeping away from the sea, and the beach-front is continually advancing as it consolidates material from the sea, it has been inferred that a lengthy period of time, not less than 200 or 300 years, had elapsed after the vessel was beached and before the wreck was noticed in 1836. Conjecture as to the origin of the ship has tended to the belief that either she was a Portuguese or Dutch vessel that had run its easting down some years prior to Hendrick Brouwer, or alternatively that she was a Spanish vessel, or a buccaneer. 'If this conjecture is correct' the *Australian Encyclopedia* muses, 'the Mahogany ship was stranded in Australia not less than 400 years ago, and her survivors, if any, may have been the first European discoverers and colonists of Terra Australis Incognita'.[18]

But it is extremely doubtful that they were. Speculation about a sixteenth century wreck is based on a child's description of an old wreck, a captain's description of something that looked like a lighter, and the flimsy theory of the sand hills moving at a certain slow rate. No fifteenth or sixteenth century artefacts have been found. It seems most likely that a vessel was wrecked there between 1788, when the settlement on the east coast began to attract regular intercourse and the whaling industry was emerging, and 1836. It is unfortunate that so much speculative energy has been devoted to the search for the Mahogany Ship in Victoria while the potential of the more tangible later shipwrecks on that coast has until recently been ignored entirely by all but the souvenir hunters.[19]

Other Australian States have their equivalents of the Mahogany Ship story. At the Wonnerup Estuary, on the south-west coast of Western Australia, settlers in the 1840s found what they described as an ancient wreck. The folk-lore which developed around this find concerned pirates, supposed to have used the estuary as a base, who had scuttled a captured vessel there. But recent research has indicated the wreck to be a longboat, lost from the French corvette *Geographe* in 1801.[20]

Queensland has more than its share of these legends, deriving from the sixteenth century Spanish presence in the north Pacific. Many Spanish ships were supposed to have been wrecked in Torres Strait and there have been stories of coins, jewelled sword hilts, skeletons dressed in rusting armour lying in a cave near Cooktown, and even the preposterous tale of an ancient wreck whose entire keel was made of silver![21]

In 1916 Commander C. Stevens of HMAS *Encounter* recovered two small brass cannon from the sand of an islet in Napier Broome Bay, on the far north of the West Australian coast. Nearby he found an abandoned Macassan campsite. The whereabouts of these two guns is still known; one is housed in the Naval establishment at Garden Island, Sydney, while the other is on display at the Fremantle Museum. Since their finding, various writers have speculated that the objects present a strong

One of the guns from Napier Broome Bay. (Photo: Pat Baker)

likelihood of a Portuguese discovery of Australia in the sixteenth century. The appearance of a 'rose' and crown on one of the guns has been cited as an indication that they were Portuguese, and the next logical step seemed to be the inference that a Portuguese vessel had been wrecked on the Australian coast.

But the presence of these cannon has a simple explanation. Throughout the nineteenth century, and indeed right up to the present, parties of Macassans have been visiting the shores of northern Australia to collect trepang, otherwise known as bêche-de-mer, or sea cucumber. After a diver takes the trepang from water up to 12 metres deep, the animal is boiled, dried in the sun, and smoked for export to China.

In February 1803 the surveyor Matthew Flinders met up with a party of Macassans off the north-east corner of Arnhem Land. A fleet of 60 prahus (the traditional Indonesian vessels) and a number of canoes, all belonging to the Rajah of Boni, and carrying 1,000 men, had left Macassar with the north-west monsoon 2 months previously on an expedition to the Australian coast, and the fleet was lying in different places to the westward, five or six prahus together. Pobassoo, the commodore of a section of the fleet, had made six or seven voyages from Macassar within the preceeding 20 years. The Macassans sometimes had skirmishes with the Aborigines, and the men were armed with daggers while each prahu carried muskets. Pobassoo's prahu carried two small brass cannon obtained from the Dutch.[22]

Numbers of prahus were wrecked over the years, and shore parties of Macassans were wiped out by the Aborigines. On such occasions the Macassans' equipment, sometimes including cannon, would have been left behind on or near the shore. The cannon at Napier Broome Bay, and others found along the coast, probably came from this source. Recent comparative, X-ray and chemical analysis of the guns has indicated Asian manufacture and nineteenth century use.[23] Analysis of the iron on one of the guns indicated that it had not been in the sea for any length of time, and thus does not originate from a wreck.

When were the guns left there? Archaeologist Campbell Macknight obtained radiocarbon estimations for three Macassan sites, varying from modern to some 800 years ago. However, he dismissed the radiocarbon dates in favour of historical evidence. The earliest Chinese consumption of bêche-de-mer from any area dates only from the sixteenth century, and the import trade from Southeast Asia did not begin before the late seventeenth century. These conditions provided for Macknight strong evidence against any Macassan bêche-de-mer fishermen coming to Australia before 1650, and he concludes that the industry must have begun between 1650 and 1750.[24] The guns are likely to have been left there during the late eighteenth or early nineteenth century, when the industry was in full swing. That is not to deny the possibility, or indeed strong likelihood, that seafarers from Indonesian islands made irregular contact with Australia on various occasions before the first Europeans. If they had adequate craft and seagoing abilities for the crossing over 40,000 years ago then surely there would be followers, despite the increasing length of the voyage. The finding of sites representing such contact offers an interesting challenge to maritime archaeology. Somewhat similar challenges are presented by the Wandjina paintings of the Aborigines.[25] The first step will be to examine the seabed adjacent to known Macassan campsites in search of underwater environments favourable to the preservation of sunken prahus, or any of their fittings and contents.

It is appropriate at this point to pause and review the importance of all the European contacts which were made with Australia prior to the first settlement. The gradual unveiling of the coastline by the Dutch was achieved principally during the years 1606 to 1644, commencing with Willem Jansz and ending with the explorer Abel Tasman. The chance discoveries of Dutch skippers following Brouwer's route gave the north, west and south coasts of Australia a fixed cartographical outline which was to remain unimproved for the next 125 years, while Tasman, by his circumnavigation, had placed the hypothetical southern continent in perspective. Certainly the Australian continent did not match the cosmographers' concept of a vast Southland embracing the southern Pacific to the extent of balancing the continents of the northern hemisphere.

Yet this was as far as their contribution went. The Dutch navigators carefully observed the outline of the coast, lest they should run ashore on their next voyage, but they were not interested in the interior, that is, the continent itself. Willem Jansz had witnessed at close hand the ferocity of the Aborigines, and saw the futility of trying to establish trade relations. Abel Tasman, the best-known Dutch Southland explorer, completed the first circumnavigation without even sighting mainland Australia. The Dutch merchants had passed close to the uranium and diamonds of northern Australia and the mountains of iron ore in the North West without realising the potential of the country. And even if they had known of the vast quantities of gold that were later to be found in Western Australia they would have been unable to exploit that resource in the absence of an easily manipulated indigenous population to supply the manpower. For the Dutch were neither miners nor manufacturers, but merely traders seeking marketable goods.

Had the Portuguese, Spanish or English been the first to discover Australia, the consequences would have been no different because no valuable trade or conquest was to be made there.

If the Dutch presence was relatively unimportant, then what of the Dutch shipwrecks? These wrecks are a valuable area of study because of the period and type of ship they represent. (The significance of the East Indiamen in regard to the study of shipbuilding will be dealt with in a later chapter.) As well, the cargoes illustrate the range of commodities which filled the hulls of outward-bound ships engaged in East Indies trade, and the wreck sites viewed as a whole form accurately dated time capsules—closed or discrete groups of cultural finds from a particular point in time. Thus they can be used by archaeologists as an aid in dating or identifying other sites or artefacts. Nor can the symbolic value of these representatives from the Age of Discovery be disregarded. Bearing all this in mind, the East Indiamen wrecked here during the seventeenth and eighteenth centuries are nevertheless of less significance to those interested in the history of Australia and Australians than those later vessels which came here deliberately to do business of some kind in Australia. For the Dutch were simply passers by. 'New Holland' was to become 'Australia'.

Chapter 3 REFERENCES

1. McIntyre, 1977, basing himself on the highly controversial Dieppe maps, and navigational arguments, contends that the Portuguese discovered and explored much of the continent before the Dutch. But Richardson, 1984, a linguist sees the Dieppe maps as representing Vietnam, while Ariel, 1984, a master mariner, sees problems with McIntyre's navigational theory.
2. Schilder, 1976, p. 43.
3. See Heeres, 1899, p. 4.
4. Schilder, 1976, p. 64.
5. Quoted in Green, 1977(i), p. 21.
6. Green, 1977(i), p. 21.
7. Drake Brockman, 1963.
8. Green, 1973, p. 1267.
9. Henderson, 1982.
10. Simpson, 1980, p. 14.
11. Playford, 1959, p. 19.
12. Major, 1859, p. 181.
13. Heeres, 1899, p. 13.
14. Halls, 1965, p. 3.
15. Bruijn, 1980, p. 261.
16. Major, 1859, p. 135.
17. Bruijn and Van Eyck, 1982, p. 3.
18. Grolier, 1965, Vol. 5, p. 461.
19. Anderson, 1981, pp. 46-47.
20. Henderson, 1980, p. 57.
21. Holthouse, 1976, p. 9.
22. Macknight, 1969, p. 66, and 1976, pp. 93-99.
23. Green, 1982, p. 80, and Crawford, 1969, p. 272.
24. Crawford, 1969, p. 103, and Macknight, 1969, p. 66.
25. Crawford, 1968.

4 Ships in Australian Waters After European Settlement

A COLONY IS BORN

The initial phase of the decisive British period of exploration of Australia owed nothing to the earlier Dutch discoveries, because the British looked towards the hitherto undiscovered east coast of the continent. That area had never been explored by trading vessels in the Pacific Ocean because of the prevailing winds. Vessels entering the South Pacific from the Atlantic sailed north towards the equator after rounding Cape Horn to avoid beating into the face of the Roaring Forties or striking drifting icebergs in those high latitudes.

It was a combination of special circumstances that was to draw Captain James Cook into the South Pacific in 1769–70. England's rising sea power and international commercial interests had accelerated over the years since 1750, the volume of shipping engaged in trade with North America, the West Indies and Asia doubling in a little more than two decades. The English were attracted by the idea of gaining wealth from trading with new lands. France, Britain's most important European rival, was also interested in new areas of trade and influence in the East. Bougainville visited Tahiti and other South Pacific islands in 1768. Cook was given secret instructions to search the South Pacific for the formerly discovered but imperfectly explored Terra Australis, which was still believed by some to occupy a large part of the ocean. The mid-eighteenth century was also a time of heightening interest in science, and Cook's first duty was to observe the transit of Venus on 3rd June 1769 at Tahiti. The observation of the transit of Venus through the unclouded skies of Tahiti would enable calculation of the distance of the sun from the earth.

Cook departed Plymouth in the *Endeavour* in August 1768, and sailed across the Atlantic. After rounding Cape Horn he turned north to warmer regions. When the scientific mission at Tahiti had been successfully completed Cook sailed south to look for Terra Australis, but after experiencing gales around latitude 40° south, he turned westward for New Zealand, which by charting he proved to be two islands rather than part of a great southern continent.

Then it was time to go home. Cook would have preferred to head south and then return via Cape Horn, because by this route he would have proved once and for all the existence or non-existence of a southern continent in the Pacific. But the condition of the ship was not sufficient for such an undertaking in the blustery southern waters, so he resolved to steer westward to fall in with the east coast of New Holland, and then follow the direction of the coast to the north towards the East Indies.

On 19th April 1770 Cook first sighted the east coast of Australia. He sailed northward along a coast of (seemingly) fertile, gently sloping hills until he found anchorage at Botany Bay, a few miles south of where Sydney now stands. Then he sailed along almost the entire east coast, and annexed it for Britain under the name New South Wales. While traversing the Queensland coast the *Endeavour* struck the Great Barrier Reef—the world's most extensive coral reef—and was held fast for more than a day and a night. Cook had to throw six of his cannon overboard before the *Endeavour* floated free. If the ship had been lost it is doubtful that the British would have established a colony in Australia in the 1780s.

Australia as described by Cook and his botanist, Joseph Banks, was not a suitable place for a colony of trade—the mainland had neither saleable commodities nor any substantial market for European goods. Nevertheless, disagreements between Atlantic seaboard countries during the years after Cook's return gave Britain an interest in this new continent.

Traditionally, historians have seen an over-supply of convicts as the only substantial reason for this interest. The North American colonists had revolted by 1776, depriving Britain of its outlet for excess convicts. About 1,000 criminals a year had been transported to the British colonies between 1750 and 1775, so after the American revolution British gaols overflowed and the walls of the prison hulks in the Thames bulged with long-term criminals permanently housed below decks. King George III announced to parliament in January 1787 that a plan had been made to colonise Australia as a British gaol, and the colony was founded when Captain Arthur Phillip's convoy of 11 convict-laden transports and their two naval consorts sailed into Botany Bay in January 1788.[1]

In recent years historians have argued for a modification of this view of Australia's origins: the settlement was intended as a base on a new and, in wartime, safer route to China than that via India and the Straits. As well, the convicts had to be got out of England and ought to be made useful. The British had also hoped to export flax and pine from Norfolk Island, off the east coast of Australia, as naval supplies for British vessels in the Indian Ocean.[2]

The principal ship of war attached to the First Fleet was the 20-gun HMS *Sirius*, a ship of about 540 tons. The *Sirus* had been built on the Thames in 1780 as the *Berwick*, and was designed for trading in the East Indies. But in loading her first cargo the vessel had caught fire and was burnt to the wales. The British Government needed a roomy ship for sending stores abroad so the Navy purchased and repaired the hull in 1781. Then in 1787, when a consort vessel was required for the convict fleet to New South Wales, the *Berwick* was re-named *Sirius*, refitted, commissioned, armed and provisioned for the voyage.

The 548 male, and 188 female convicts who survived the First Fleet voyage (40 having died *en route*) were discharged at Sydney Cove. New South Wales then became the most extensive penal colony on earth. Several weeks later the brig

Captain John Hunter, who lost the *Sirius*.

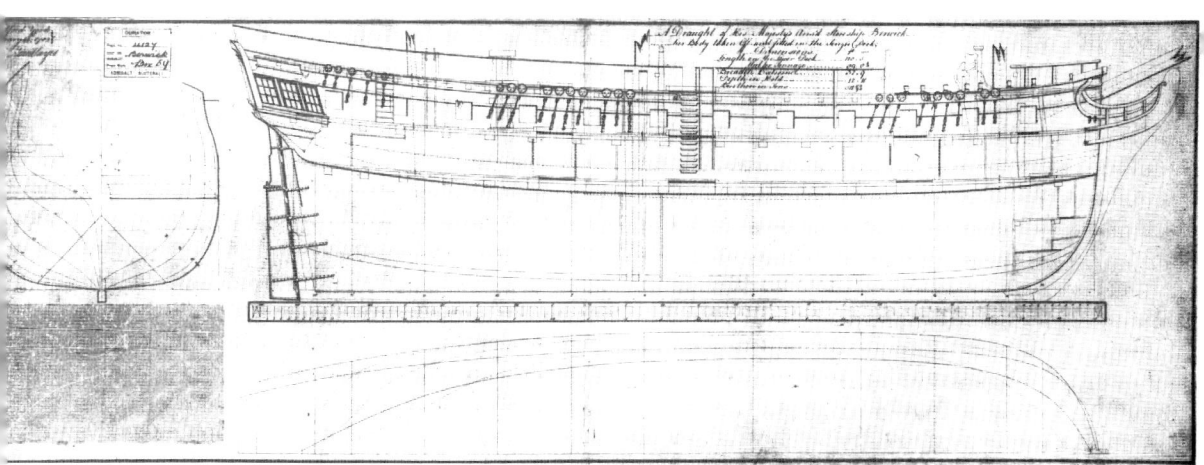

Admiralty draught of the *Sirius* (or *Berwick*). (Photo: State Library of New South Wales)

Supply took a small party to Norfolk Island to prevent it being occupied by any other European power.

The unfamiliar environment soon made survival the aim of the convict colonies. Because of their extreme isolation the convicts tried desperately to coax the unwilling soil to produce. For as the months passed no food ships arrived from Britain to replenish their faltering stocks. In June 1788 the Lieutenant Governor, fearful of general starvation, sent the *Supply* some 650 kilometres to Lord Howe Island on an unsuccessful turtle hunt. Then in October he sent his largest ship, the *Sirius*, to Cape Town for supplies.

The shortest route for Captain John Hunter would have been around Tasmania and then west to the Cape of Good Hope, an 8,000-kilometre passage. Hunter saw the disadvantages of this route: he would have to beat into the prevailing westerlies all the way across the Great Australian Bight and then across the vast expanse of the southern Indian Ocean. For this reason, and because the *Sirius* developed a leak, he sailed east instead of west and had favourable winds the whole way. By sailing past Cape Horn he added some 6,500 kilometres to his passage but still saved time. Despite the ice and scurvy, he raced from Cape Horn to Cape of Good Hope where he bought wheat and barley before continuing eastward and home to Sydney. The *Sirius* had sailed right around the world and been away only 7 months. Yet she carried unhappy tidings: the news in Cape Town suggested that no relief ships had left England.[3]

By 1790 the colony had been reduced to half rations so the *Sirius* was sent with the *Supply* to Norfolk Island, which had better resources for food gatherers. The two vessels had on board two companies of marines consisting of 65 officers and men with 5 women and children, and 116 male and 67 female convicts with 27 children, all of whom were to be re-established on Norfolk Island. They made the island on 13th March and landed all the people the next day. Then bad weather came in, preventing any provisions being landed until the 19th when the weather turned fine and the *Sirius* was brought into Sydney Bay, close to the settlement of Kingston.

The boats were employed landing provisions when the ship was seen to be drifting fast to leeward. Hunter made sail to get out of the bay, but the wind shifted and the ship missed stays, so that although an anchor was let go she struck and was wrecked on the reef. One contemporary writer observed 'her bottom bulged immediately, and the masts were soon cut away, and the gallant ship upon which hung the hopes of the colony was now a complete wreck'.[4] It was hoped that the greatest part of the provisions would be saved, but many of the officers lost their effects.

On the loss of the *Sirius* Norfolk Island was put under martial law. Two convicts, John Branagan and William Dring, volunteered to board the wreck to offload livestock, but upon reaching the wreck through the surf they refused to return and set fire to the *Sirius*. Some volleys of small arms were fired, and then a three-pounder cannon blast was directed at the ship, without having the desired effect of persuading the recalcitrants to leave. Soon afterwards John Arscott, a carpenter, went on board and forced the two to leave the ship by a hawser. Arscott extinguished the fire, but not before it had burnt through the gun deck.

On the 28th of March a very heavy surf forced the wreck broadside further in upon the reef. The ballast had fallen out through the bottom of the hull, making it lighter, and the wreck was thrown more than its own length towards the shore. In this new position the wreck was almost out of the reach of the surf, so the crew were able to salvage provisions, fittings, and even the majority of the ship's guns.

It was a sad irony: the first ship to be wrecked in Australian waters after the arrival of the First Fleet was the very ship that had led that fleet out from Britain in 1788.

In the Sydney penitentiary the news of the loss of the *Sirius* had an immediate and decisive effect on morale. An officer, upon hearing the news, wrote:

> The country, my Lord, is past all dispute a wretched one, very wretched, and totally incapable of yielding to Great Britain a return for colonising it. There is no likelihood that the colony will be able to support itself in grain or animal food for many years to come, so that a regular annual expense is entailed on the mother country as long as it shall be kept.[5]

It was now necessary to drastically reduce the rations. Most labour was laid aside because the men were too weak, and all the boats were employed in fishing to build up food stocks.

The news of the *Sirius*'s loss preceded by only a few weeks the agonising information that the *Guardian*, a fast naval vessel bringing more than a year's provisions from England, had been lost *en route* in the southern Indian Ocean. John Williams, the boatswain, related his terrible experience to his agents in London:

> December the 22, in the latt. of 42.52 south, long. 42.22 east, we fell in with several islands of ice. The ice was so lofty that itt drifted faster than we expected, by the wind having so much hold of itt, the weather being so thick that wee could scarce see the length of the ship, the wind blowing fresh that we run foul of it and received a great deale of dammage, knocked away our rudder, broake the tiller in three pieces, broake one of the after beams in two, knocked the sternpost from the keel, and dammaged the ship in a shocking manner. We was about six hundred leagues from any land. There was about fifty six men missing; a number drowned jumping into the boats; the sea ran so high that the boats could scarce live. The commander had a strong resolution, for he said he would soner go down in the ship than he wold quid hur. All the officers left in the ship is the commander, the carpenter, one midshipman, and myself. We got up a new forecourse and stuck itt full of oakum and rags, and put itt under the ship's bottom; this called fothering the ship. We found some benefit by itt, for pumping and bailing we gained on hur; that gave us some hopse of saving our lives. We was in this terable situation for nine weeks before we got to the Cape of Good Hope. Sometimes our upper deck scuppers was under water outside, and the ship leying like a log on the water, and the sea breaking over her as if she was a rock in the sea. Sixteen feet of water was the common run for the nine weeks in the hold . . .[6]

The *Guardian*, under the command of Lieutenant Riou, eventually limped into

False Bay, South Africa, only to be driven ashore and totally wrecked in a fierce gale on 12th April 1780. The combined expenses on account of the *Sirius* and *Guardian* amounted to £73,917. This was a very substantial loss when it is considered that the total expenses and charges of the entire civil and military establishments, from the beginning of the colony until October 1792, was only £473,000. But the *Lady Juliana*, which brought the news of the *Guardian* tragedy to Sydney, carried supplies and glad tidings: the famine was over and the second fleet was soon to arrive. There was an enlarged plan for the colony, in which free settlers were to play a part.

While the *Guardian* had been refreshing at the Cape, prior to striking the iceberg, her commander, Lieutenant Riou, met Lieutenant William Bligh, who had lost his ship and was returning to England.

Bligh had sailed with Captain Cook on the last of his voyages of discovery in the Pacific and had been an eye witness to Cook's death. In 1787 he was given the command of an armed transport, HMS *Bounty*, for an expedition to Tahiti to gather breadfruit trees. These were to be taken to the West Indies sugar plantations as a potential food supply for the slaves.

Bligh arrived at the island toward the end of 1788, but had to wait there for 5 months until the seedlings were mature enough to be planted in pots and taken on board. During that time the *Bounty* crew formed strong attachments with Tahitian women.[7]

The plants were eventually loaded and the *Bounty* left Tahiti on the 4th of April 1789. Three weeks later, just as dawn was breaking off the island of Tofua, Bligh was rudely awakened. Fletcher Christian, the senior master's mate, accompanied by three seamen, forced him on deck. The crew of the *Bounty* had mutinied, exasperated by Bligh's harsh discipline and allured by the promise of a life of dissipation on the beautiful islands they had left behind. They set Bligh and 18 loyal men adrift in an open boat. But with great determination and skill, Bligh sailed the 7-metre boat through Torres Strait to safety on Timor. When the ordeal was over, the exhausted men had sailed 3,618 nautical miles (some 5,800 kilometres) in 41 days without the loss of a single life.[8]

When Bligh's dispatch with news of the mutiny reached the Admiralty only one course could be taken. Mutiny was anathema to the eighteenth century British navy, and exemplary punishment was the only thing for mutineers. A larger and more powerful ship would have to be sent to deal with the errant ship and seamen.

The British Government vaguely considered sending HMS *Sirius* from Port Jackson, but that plan was not acted upon. Instead the 24-gun HMS *Pandora* went out late in 1790 under the command of Captain Edward Edwards to search for the mutineers and make another attempt to bring back breadfruit trees. Fourteen men were captured at Tahiti. They told Edwards that Fletcher Christian, with eight of the mutinous crew and some Tahitians, had sailed for an unknown destination. Their refuge was in fact Pitcairn Island, 2,100 kilometres south east of Tahiti, where the mutineers burned the *Bounty* ashore in January 1790. It was not until 18 years after the mutiny, when the Boston whaler *Topaz* touched at the island, that the mutineers'

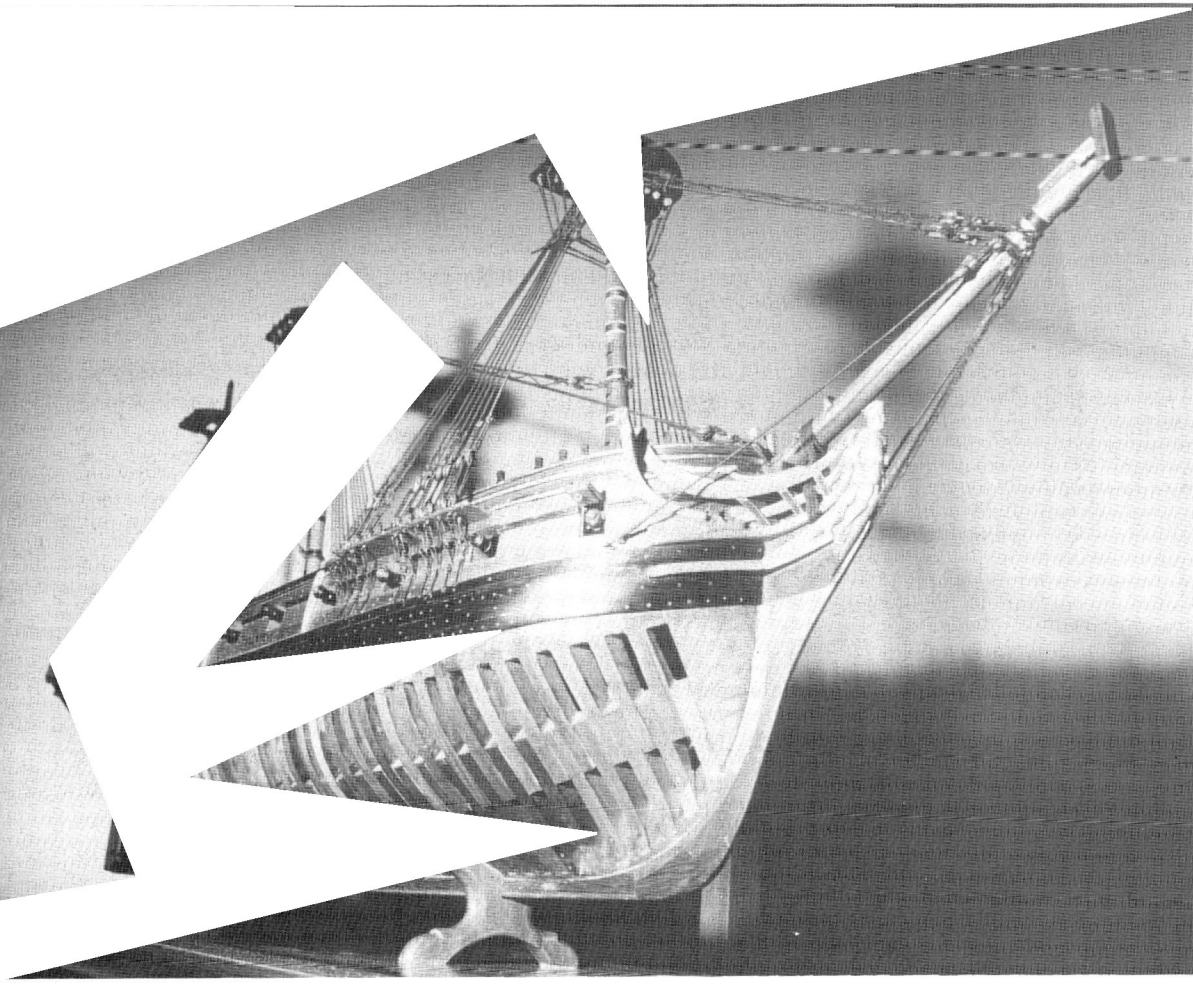

A model of HMS *Pandora*, by Keith Hobbs. (Photo: Pat Baker)

refuge was finally found. Only one mutineer, John Adams, had survived the quarrels with the Tahitians.

Captain Edwards never did find these mutineers, and his only option was to set a course for home, with some of the *Bounty* mutineers and a load of breadfruit plants. The prisoners were confined in irons in a specially built wooden cell, 3.35 by 5.49 metres, situated on the quarterdeck and referred to as 'Pandora's Box'.[9] The entrance was a 0.51-metre square scuttle on top, which was bolted down at all times.

On 28th August 1791 the *Pandora* was approaching Torres Strait. Edwards sent the yawl to examine an opening in the Great Barrier Reef. At 5 p.m. the boat crew signalled that a passage had been found. Night was coming on so they were ordered to return to the ship. At that time the soundings gave 200 metres. By 7 p.m., however, the lead line gave only 90 metres. Suddenly the vessel struck so hard on the reef that with every surge it appeared to the crew that the masts might come crashing down.

Soon there were 3 metres of water in the hold and the prisoners, fearful that the ship was going down, broke their irons in readiness. However, the unfeeling Captain Edwards ordered them to be handcuffed and leg-ironed again with all the iron that could be mustered, while sentinels were ordered to fire into the box if the prisoners moved. Captain Edwards's conduct can be understood, if not condoned, if he is considered in the context of his time. Earlier in the year Britain had nearly gone to war with Spain. The French Revolution was in full swing on the other side of the Channel. International tension was high and social tension even more so. Ideas of equality were presenting a real threat to established authority. The result was that the Royal Navy, Britain's main line of defence, could not afford a show of weakness under any circumstances.[10]

The crew threw some of the guns overboard to lighten the ship, but during the night several of the pumps broke down and it became clear that the ship could not be saved. As the *Pandora* went down, Captain Edwards leaped from the stern and swam to the pinnace. The boson's mate threw the scuttle overboard, allowing those of the prisoners who could rid themselves of their hand and leg-irons to scramble through the opening in their cell and struggle to the surface for air. Four mutineers drowned together with 31 of the ship's company. The *Pandora* was irretrievably lost in 30 metres of water.

The survivors camped on a small island nearby for 19 days. Then they set out for Batavia, which they reached safely. Arriving back in England in June 1792, six of the ten surviving mutineers were sentenced to death. The incompetent Edwards was court martialled for sailing the *Pandora* on to the reef, but he was not convicted. Meanwhile, Bligh, with two more ships, had already left England for Tahiti in another attempt to gather breadfruit plants. This time he was successful, but ironically the breadfruit was shunned by the slaves in Jamaica. The project had been an even greater failure than the colony in Australia at that time appeared to be.

Communications between England and Australia during the early years of settlement were inevitably erratic because of the great distance between the two countries. In this situation, Asian commerce could be expected to develop and fill the gaps. Initially, overseas trade was severely restricted by the monopoly given to the East India Company under the Navigation Acts. The loss of the *Guardian* in 1790 led certain Indian merchants to offer to supply the colony with provisions, and as the East India Company did not object to this, a regular intercourse developed during the 1790s. Cargoes from Bengal fed and equipped the colony but, as the historian Geoffrey Blainey has observed, they 'also gave it a hangover'.[11] Bengal rum was Australia's first national drink and at times was also a form of national currency.

The officers of the New South Wales Corps were the group best equipped to engage in this trade. During the 1790s, despite the efforts of successive colonial governors, they developed a trading monopoly in which rum was the established medium of exchange. The rum was monopolised to raise its price, while its consumption was pushed to the limit to allow the monopolists to make huge profits. When William Bligh of *Bounty* fame became Governor of New South Wales he attempted to destroy the trade in rum, but was arrested by those same officers in what became known as the 'Rum Rebellion'.

One of the rum-trade vessels of the 1790s was the ship *Sydney Cove*. This vessel, under the command of Captain Guy Hamilton, left Bengal on the 10th November 1796 on a speculative voyage to Port Jackson, New South Wales. She was laden with 7,000 gallons of spirits and a quantity of general merchandise. The vessel was sent out by the merchants Campbell, Clarke and Co. who had a distillery in India.

On 13th December, in latitude 15° 30 ', a little south of Cocos Island in the Indian Ocean, the ship experienced a severe gale. High seas and strong squalls continued for the next 4 weeks as the *Sydney Cove* ploughed southwards past the west coast of the Australian continent, and then turned eastwards towards the southern tip of Tasmania.

Captain Hamilton in his account tells us that the ship laboured a good deal and made 0.15 to 0.2 metres of water every hour from a leak under the starboard bow. On the 13th January a thrummed sail (sewn over with rope yarn) was passed under the ship's bottom, reducing the leak. Then on the 25th January a gale of extreme violence set in from the south west. In furling the topsails the second mate was lost from the main topsail yard-arm. The wind blew with such fury that a new foresail, a main topsail and the driver were torn from the yards even though they had been furled and secured. The ship was then hove to, having no sails bent to keep her ahead of the sea.

The gale continued and the leak got worse. The Indian crewmen suffered badly from the intense cold and constant rain, and finally Captain Hamilton found that neither pleas nor orders could prevail upon them to work on deck at the pumps. So all hands were sent down to bale from the well, and with diligence they reduced the level from 1.2 to 0.6 metres. There was no intermission from pumping and baling: two men dropped dead under the labour, and a third died a few hours afterwards. After a severe and tedious night the well was cleared of water, and as the gale had abated a little a new foresail was bent, and the ship made sail with the wind at the west.

On the 27th January a new piece of canvas was taken under the bows, reducing the leak from 0.3 to 0.2 metres. The vessel moved eastward past the south coast of Tasmania (mariners at that time being unaware of the existence of Bass Strait) and then headed north, sighting Maria Island on the 4th. But from this time onwards, things got worse:

> On the 8th of February, observed in the latitude 40° 56 ' south, longitude by timepiece, 149° 40 ' east of Greenwich. The gale now increased to a perfect hurricane, with a dreadful sea. At half-past 3 p.m. sprung a new leak, which gained so fast on the pumps as rendered it necessary to bear up for land to save

the lives of the people, and if possible, to get the ship into a place of security. Bore in accordingly for the land and made more sail, Cape Barras by accounts W. ½ N. or W. and by N., distant by accounts 90 miles. The cargo was thrown overboard; but notwithstanding every exertion to keep the leak under, at 5 p.m. there were 2½ feet water in the well, and hourly gaining. At 8 p.m. the water had increased to 5 feet, and the ship settling fast, the longboat was got clear—still running west and carrying a press of sail in order to get in with the land. By midnight the water was nearly up to the lower deck hatches. At half past 12 saw the land about 2 miles distant: but appearing to be high perpendicular rocks with a heavy surf, it was thought advisable to heave to till the morning. At daybreak the water was over the coamings of the lower deck hatches, and the vessel lying on one side with the channels on the water. At daylight, having with difficulty got her head round, made all sail possible towards the land, but from her being so much water-logged she would hardly answer her helm.[12]

The *Sydney Cove* stood in for Preservation Island in Bass Strait, pressing on until she struck on a sandy bottom in 5.7 metres of water. The crew were safely landed on the island and 3 weeks later the longboat, manned by 17 of the best of the crew, set out for Port Jackson to seek help. The longboat was wrecked on the mainland coast some 350 kilometres to the southward of Port Jackson, but all aboard got ashore and travelled along the coast. Fatigue and attacks by the Aborigines reduced the number of survivors from 15 to 3 before they reached the settlement on 15th May. As they trudged north these men discovered the first commercially viable Australian coal, at a place called Coalcliff.

Soon afterwards the 42-ton Colonial Schooner *Francis* was despatched for Preservation Island in company with the sloop-rigged, decked longboat *Eliza* of 10 tons. At the island the two vessels were loaded with cargo salvaged from the wreck, and set upon a course for Port Jackson, carrying with them the remainder of the *Sydney Cove*'s crew. However, the *Eliza* was wrecked *en route*, and none of the men on board ever seen again. A wreck found near Port Phillip in 1803 was suggested as possibly being the *Eliza*.[13]

A few months after the *Sydney Cove* had been wrecked Governor Hunter (the former commander of the *Sirius*) sent George Bass southwards in a whaleboat. He discovered Bass Strait and proved for the first time that Van Diemen's Land was a separate island. The opening of Bass Strait shortened the route to Sydney from the west by some 1,100 kilometres, and had a marked influence on the development of maritime trade to Australia.

Extensive salvage of the *Sydney Cove* took place soon after the vessel sank. When the rescue vessels *Francis* and *Eliza* had arrived at Preservation Island in June 1797 they took on full cargoes of material salvaged from the *Sydney Cove*, and left six volunteers behind in charge of the remainder. These men lived in a house they had built on the island from materials salvaged from the wreck.

The *Francis* returned for a second load in December 1797, and took on 3,500 gallons of rum and a mare. The rum was sold at such enormous prices in Sydney

This carved slate, found on Preservation Island by lessee Bruce Bensemann, is thought to represent the *Sydney Cove*. (Photo: Paul Clark, National Parks and Wildlife Service, Tasmania)

that one observer thought the owners would not have lost much from the wreck. Even 'common' cups and saucers salvaged sold for 22 shillings each.[14]

The *Francis* sailed a third time for Preservation Island in February 1798. On board was Lieutenant Matthew Flinders who wanted to conduct nautical observations. Flinders anchored at Preservation Island and found remnants of the *Sydney Cove* and its cargo scattered by westerly gales. He also noted the presence of extensive herds of fur seals, which were soon to become the focus of a valuable export industry. Ironically, the sealing industry in turn was responsible for the survival of the Tasmanian Aborigines. The women taken to the small islands by sealers had a recognised role to play in that industry and thus escaped the fate of Truganini and her fellows who were all either killed by the Europeans or rounded up and transported to unfamiliar localities to linger and die. The development of maritime industries in Bass Strait led indirectly to further depredations upon the wreck. In 1804 Governor King expressed concern about an American ship working in the area, whose crew were building a vessel from the *Sydney Cove* wreck, and had erected a dwelling from the remains. The colony at that time remained a convict settlement and King wanted to prevent the building of any craft which might be used by escapees. He directed that if the Americans did not cease their boat-building activity he would have the King's mark put on the timbers.

What then of the overall significance of Australia's eighteenth century post-European-settlement shipwrecks? Discounting the *Guardian*, which sank off South Africa, the 10-ton sloop *Eliza* and the longboat lost from the *Sydney Cove* (which was so small that it is doubtful that any part of it would still survive), there are only three remaining—HMS *Sirius*, HMS *Pandora* and the *Sydney Cove*. The *Sirius*, principal warship attached to the First Fleet, was a vessel of crucial importance to Australia during its foundation years. After the First Fleet had arrived at Port Jackson the *Sirius* continued its role of protector, and also assumed the role of provider. When the *Sirius* was lost on the reef at Norfolk Island the news had a devastating effect upon the morale of the mainland colony. It was the first known 'Australian' shipwreck, in the sense of a ship lost while being used by a people living in Australia. It is true that the ship was wrecked on a shallow, turbulent reef and has broken up as much as the East Indiamen which had earlier crashed on to the reefs off the west coast of Australia. The substantial salvage undertaken soon after the ship was wrecked somewhat reduces the archaeological potential of the remains. Nevertheless, study of the wreck of the *Sirius*, in conjunction with the available historical data, may be expected to increase understanding of the events occurring on Norfolk Island at that crisis period, and to expand knowledge of the ship itself.

The *Pandora*'s voyage and wreck has less direct relevance to the foundation years of Australian history, but the British presence in Tahiti and Port Jackson at that time were both to some extent manifestations of a British desire for naval influence in the region of the South Pacific. The *Pandora* excavation is providing a dramatic illustration of aspects of an episode in the story of the mutiny on the *Bounty*. The state of preservation of the wreck is such that it is proving to be the most interesting maritime archaeological site yet excavated in Australian waters.

The *Sydney Cove*, although less spectacular, has much to offer. This was the first post-settlement wreck of a merchant ship in Australian waters. It was the development of trade which enabled Australia to move from its limited status as a convict outpost to become a colony of settlement, and eventually to become a nation. Rum, the cargo of the *Sydney Cove*, was at the centre of the social, economic and political changes occurring at that time.

EARLY DEVELOPMENT, 1800–1850

Whereas only five substantial vessels (including two Dutchmen) are known to have been wrecked in Australian waters during the eighteenth century, several thousands were lost during the nineteenth century. It would be pointless to attempt to describe here the circumstances surrounding every one of these casualties, but some observations can be made about trends. Looking at the first half of the nineteenth century, it is useful to consider the listing of shipwrecks which has been made by Charles Bateson in his *Australian Shipwrecks*.[15] I will make use of the information in that volume to examine some statistical groupings of shipwrecks in that period.

Ignoring vessels which were lost outside of Australian waters, or which were not total losses, the types of substantial vessels wrecked can be divided up approximately as follows: 110 schooners, 55 cutters, 42 brigs, 38 ships, 35 barques, 33 sloops, 4 brigantines and 4 steamers. There was also a scattering of small vessels such as ketches, smacks, dandies and riverboats. Terminology used for different types of craft varied over time, occupation, and region, but some characteristics can be ascribed to the major types in Australian waters during this period.

The schooner was at that time a smallish seagoing fore-and-aft-rigged vessel. It originally had two masts but later some stood three or more, and carried one or more topsails. The average tonnage of the schooners wrecked during the period was about 50 tons, but they ranged from 10 to 200 tons. The smallest schooners were limited to work on sheltered coastal runs, but the larger ones were employed in international trade and regularly plied the Pacific and Asian waters.

The sloop was a small one-masted fore-and-aft-rigged vessel with a standing bowsprit. It was very similar to the cutter; a small single-masted vessel furnished with a straight running bowsprit. Both rigs were generally used for inshore work, but some of the larger cutters were regularly sent on inter-colonial voyages. The sloops wrecked during the period 1800 to 1850 averaged about 20 tons and ranged from 6 to 64 tons. These small vessels were most common during the first quarter of the century. The cutters wrecked averaged about 25 tons and ranged from 7 to 66 tons.

Brigs were two-masted square-rigged vessels, carrying also on the main mast a lower fore-and-aft sail with a gaff and boom. The brigs wrecked during the period averaged 180 tons but ranged from 70 to 351 tons. This handy rig was employed on all manner of deepsea work, including regular voyages to Asia and Europe.

The ship was a large sea-going square-rigged vessel having a bowsprit and three or more masts, each of which consisted of a lower, top and topgallant mast. The size of the ships wrecked between 1800 and 1850 averaged about 525 tons and ranged from 242 to 710 tons. At the beginning of the nineteenth century these vessels carried out the bulk of the long-distance carrying, but by 1850 this rig had been successfully challenged by that of the economical barque. The barque was a three- (or more) masted vessel with the fore and main masts square rigged and the mizzen fore-and-aft rigged. The barques wrecked during the period averaged 310 tons and ranged from 150 to 518 tons.

Paddlewheel steamers were introduced for passengers and cargo transport on rivers and between major east coast ports during the second quarter of the nine-

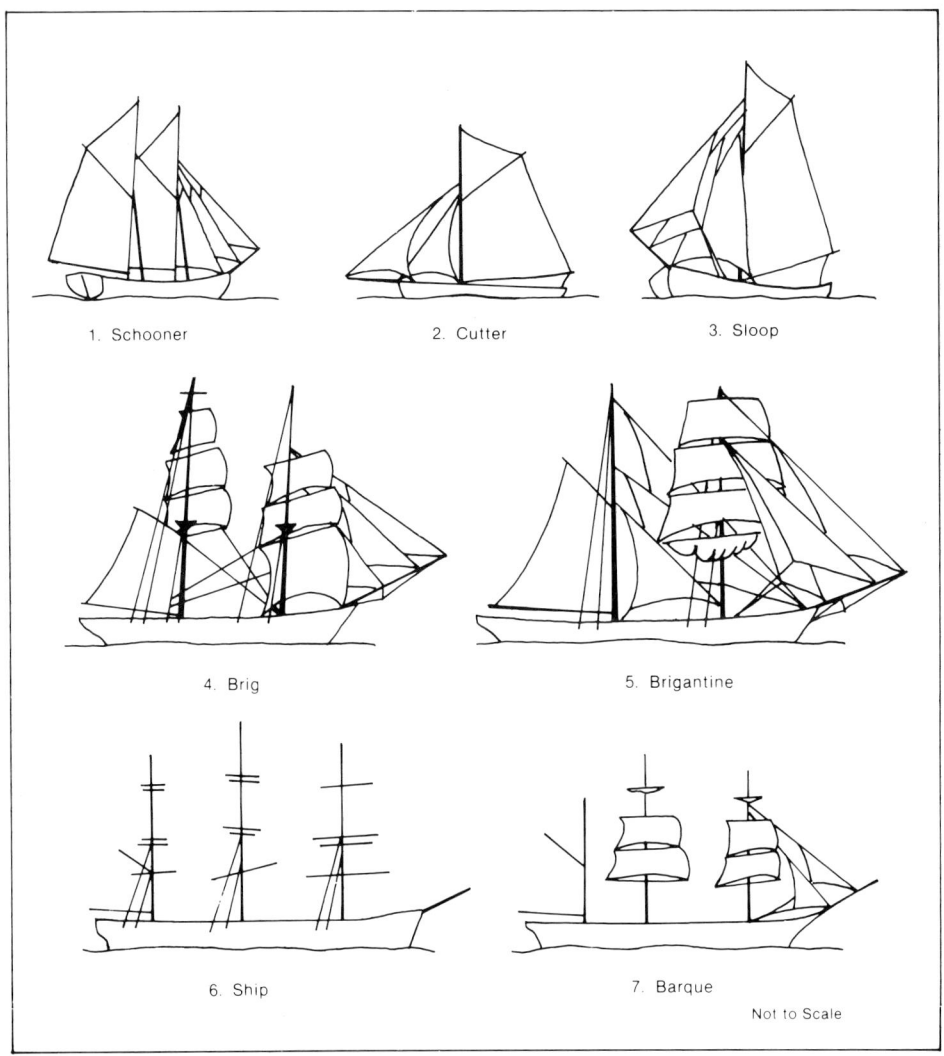

Sailing vessel rigs, taken from Alan Moore's *Mast and Sail*.

teenth century. Steamships were then in a relatively primitive stage of development and on the sparsely populated Australian coast sometimes proved to be economically unsound. Up to 1850 the average tonnage of the steamers wrecked was about 185 tons, and they ranged from 81 tons to 297 tons.

The heaviest concentrations of wrecks during the early nineteenth century were along the New South Wales coast, Torres Strait and Bass Strait. The wrecks along the New South Wales coast, particularly in the vicinity of such large ports as Sydney and Newcastle, may be explained by the density of both coastal and overseas traffic in that area. In Bass Strait many vessels were wrecked because of the weather conditions, while in Torres Strait the problem was to find a clear passage through the tangle of shallow reefs. Most of the traffic bound for Sydney from the west, or bound for the west from Sydney, used one of these straits.

While the large square-rigged vessels tended to be used in long distance ocean trades, and the small fore-and-aft craft generally plied the inshore coastal runs, there was a certain amount of overlap, depending on the industry in which the particular vessel was employed. The function of a vessel determined to some extent not only its rig, but also the way in which that vessel was fitted out, and the cargo it carried.

Archaeologists interested in the general development of Australia may well find that the cargoes of these vessels are more interesting than the hulls of the ships themselves, since many aspects of early nineteenth century shipbuilding are well documented, whereas relatively little work has been done on Australian trade over the same time.

In some cases, however, the hull itself is more interesting than the cargo. If a specialised vessel was re-registered as a general trading vessel and then wrecked, the hull of that vessel might be worthy of study because of what it had been—whether a warship, pirate, slaver or whatever. The snow brig *James Matthews*, wrecked near Fremantle in 1841, was an ex-slaver. Before her capture by the British Navy in 1837 she had carried slaves from Africa to the West Indies. The hull of the *James Matthews* provides the opportunity to study the design of a ship involved in this forbidden trade.

The American China trader *Rapid* provides a similar example. She left Boston for Canton in September 1810 and, after sailing across the Indian Ocean at a faster rate of knots than her captain realised, struck the Australian coast at night a little south of North West Cape. Excavation of that site is revealing information about an important period of development in American shipbuilding.

Many of the smaller traders which operated between the Australian colonies were built in Australia. The study of some of these hulls would throw light on how the Australian shipbuilders adapted to the new conditions and requirements in the face of variable supplies of fittings and the new, unfamiliar timbers available.

The scheme for a convict settlement at Port Jackson required the chartering of a number of convict transport vessels. These were ordinary British merchantmen. Vessels are not known to have been specially designed and built as convict ships. But it was desirable that the vessels selected have plenty of head room between decks, and they needed fitting out with heavy bulwark divisions to maintain the separation of prisoners from their guards, or, on the vessels carrying female convicts, separation of prisoners from the crew.

The convicts, once landed in Australia, had periodic requirements for supplies, which were brought out by sea from England and Asia on large sturdy English East Indiamen. After unloading in the Colonies, these transports and storeships would either seek return cargoes in Australia or Asia, or go whaling.[16]

The first convict transport to be lost in Australian waters went down with a bang, and observers thought it might take half of Sydney town with it. Completing her maiden voyage from London and Ireland with 209 male prisoners, the ship *Three Bees* arrived at Sydney on 6th May 1814.

The prisoners had disembarked for some days when at about 5 o'clock on the evening of 20th May a fire was discovered on board. The flames quickly assumed control and all efforts to extinguish them were rendered completely useless. The crew were doubly concerned because a very large quantity of gunpowder was known to be situated immediately adjacent to the seat of the blaze. There was no alternative but to immediately abandon ship. A witness observed:

> At this crisis, little short of the total destruction of the town of Sydney was expected every moment to take place by the explosion of the magazine. The alarm was so great that numbers of the inhabitants deserted their houses, and fled into the country to avoid being buried in its ruins. Fourteen guns, some loaded with ball and some with grape shot, exploded, sending their contents in various directions as the ship drifted, through the town, fortunately however without doing any damage further than the breaking a window in the Naval Officer's house and shattering a writing desk that lay within it. At this time a light breeze blowing off the shore, and the cable being cut, the vessel drifted to the extremity of the cove where she struck on some projecting rocks called Bennelong's Point, and here the expected explosion took place.[17]

When it came, the explosion was by no means as tremendous as the terrified residents of Sydney had thought it would be, perhaps due to some previous wetting of the gunpowder. The fine ship was soon burnt down to the water's edge, at the spot where the Opera House now stands.

Another convict ship was lost in 1817. The *Fame* had brought out 120 male prisoners from Britain to Sydney. After unloading the convicts, she took on a cargo of horses and set out for Batavia and Calcutta on her way home to England. But on her way northward around Cape York she was wrecked in Torres Strait. The *Fame* was the first of a number of returning convict ships to be wrecked off Queensland. In March 1825 the ship *Henry*, having delivered female convicts at Hobart, departed Sydney for Batavia, and was wrecked in Torres Strait. During the same year the *Royal Charlotte* left 136 male prisoners at Port Jackson and sailed for Batavia and Calcutta. She was wrecked on the Frederick Reef, 465 kilometres east of the Queensland town of Mackay. Similarly, the *Prince Regent* was wrecked on the Great Barrier Reef in 1827, and the *Governor Ready* in Torres Strait in 1829.

All these vessels were returning convict transports. Before they were wrecked the convicts with their chains had disembarked, and the special erections in the holds dismantled. But in 1835 two convict ships, both outward bound from Britain to Van Diemens Land or New South Wales and crammed with prisoners, were lost near Tasmania.

The *George III* was totally wrecked in April 1835 in one of the approaches to Hobart, with the loss of 131 lives. The vessel had struck an uncharted rock and 127 prisoners were drowned in their cells below deck as the water rose. All the guards, most of the crew and 81 convicts were saved in the ship's boats and a rescue vessel.

A representation of the *Edwin Fox* as a convict transport in 1858, by Ross Shardlow.

A month later an even worse tragedy took place in Bass Strait. The *Neva*, bound for Port Jackson with 150 women convicts, 55 children and 9 free women, was totally wrecked on King Island with a loss of 218 lives. Attempts were made to launch the boats but they were dashed to pieces or capsised. Many of the women in the cabin were killed or injured when the poop deck fell in on them.

These and other outward-bound (from Britain) convict ships lost on the Australian coast provide the potential for a fascinating study of the conditions under which the prisoners suffered during the outward voyage. If such a wreck was found in good condition it might provide information about the space allocated to each convict, the means of confining the convicts, the equipment used for maintaining order, and the living space allocated to the guards.

The *Edwin Fox* survives, if only just, as an example of a vessel used as an Australian convict ship. The vessel was built of teak in Bengal in 1853 (one of the last ships constructed for the East India Company) and initially engaged in the tea trade. In 1858 she transported 280 political prisoners as convicts to Fremantle, but it is certain that the fittings related to that voyage would have been subsequently removed. Until recently her hull was still floating at Picton, and New Zealanders hope that the one-time convict transport will someday be fully restored, either in New Zealand or Australia.

The mere presence of shipping tonnage was a stimulus towards the production of export commodities. By 1800 a substantial proportion of the 5,000 inhabitants of Sydney no longer had the right to draw upon the Government stores for their food and other requirements. It became necessary for these men to take to the sea in search of seal skins and sea elephant oil for the Chinese, English and Indian markets. They brought coals and cedar from Newcastle for shipment to the Cape of Good Hope and India; pork from Tahiti to supplement the rations in Sydney; they sought sandalwood and bêche-de-mer on the Barrier Reef and in Fiji, Tonga and the Marquesas for shipment to China. Wheat, oil, seal skins and kangaroo hides came from Tasmania, and flax, potatoes and timber from New Zealand.

Upon the arrival of the First Fleet in Australia and the establishment of the convict

The *Edwin Fox* lay as a hulk at Picton in the 1950s, 100 years later.

colony, the need was immediately apparent for sufficient means of sea transport and communications. The convict transports left for other destinations, leaving behind in the colony just two vessels belonging to the British government; the *Sirius* and her tender *Supply*. Restrictions were placed on the private building or importing of small craft to best ensure that the island-continent penitentiary functioned as an escape-proof institution. But there was an urgent and continuing need for vessels to perform a wide range of tasks along the Australian coast, like survey missions, the transport of troops, convicts and passengers, supplying outposts with provisions, general communications, fishing for food supplies, and other work which had to be done by the colonial government in the absence of sufficient number of privately owned vessels. Thus small brigs, schooners and sloops were built, bought or otherwise acquired by the colonial government and equipped for this routine work.

The sloop *Norfolk* was built for the colonial government on Norfolk Island. She was made famous by the circumnavigation of Tasmania under command of Bass and Flinders. In October 1800 the *Norfolk* was seized by convicts at the mouth of the Hawkesbury River in New South Wales, accidentally run on shore, and bilged. Another colonial-built government vessel was the brig *Elizabeth Henrietta*, whose keel was laid down at Sydney in 1797. The *Elizabeth Henrietta*, built of ironbark and stringybark timber, was wrecked at Newcastle, New South Wales, in 1825. The government schooner *Francis* was brought out from England in frame in 1792, was launched in 1793, and was engaged in salvage of cargo from the *Sydney Cove* wreck in Bass Strait in 1797. She was herself wrecked ashore at Newcastle in 1805.

Some colonial government vessels were acquired from overseas. The brig *Lady Nelson*, built at Deptford with three sliding keels or centreboards was brought to Australia in 1800 and employed as a survey vessel until 1825, when she was captured and destroyed by natives near Timor. The cutter *Mermaid* was built at Calcutta and happened to be in Sydney on a trading voyage in 1817 when the Government required a vessel for surveys of the north coast. The *Mermaid* was wrecked on the Great Barrier Reef in 1829.

The *Estramina* was built at Callao in 1803 and was captured from the Spanish by William Campbell of the brig *Harrington*, off the coast of Peru in 1804. The colonial government of Australia then bought the vessel and employed her until she was wrecked at Newcastle in 1816. The wrecks of these government vessels, if they are found to be in good condition, might be expected to tell us something about the administration of the early Australian colonies.

Sealing was one of Australia's first maritime industries. Typically small concerns operated in isolated areas along the south coast of the continent. Because of the nature of early sealing, it may be expected that there would be many important gaps in the documentary record of this industry. Unfortunately, in many cases the circumstances surrounding the reported losses of sealing vessels seem to indicate that the wrecks have a limited potential for throwing light on the activities of the sealers.

The first sealer wrecked in Australian waters was not British, American or colonial, but French. The 90-ton schooner *L'Entreprise* left Sydney in October 1803 to go sealing, but was wrecked in Bass Strait. Her rigging, sails and other items salvaged from the wreck were auctioned in Sydney. A much larger vessel, the 248-ton ship *Campbell Macquarie*, was wrecked at Macquarie Island in 1812 after loading 1,650 sealskins, but this wreck too was heavily salvaged and later set on fire. Similarly, the brig *Belinda*, wrecked at the sealing grounds on the south-west coast of Australia in 1824, was subject to some salvage soon afterwards. In 1829 the Hobart-built schooner *Black Swan*, which had been sealing, was wrecked east of Wilson's Promontory in Victoria, but the records do not indicate whether salvage took place on the wreck.

Free colonists, and some escaped convicts, worked in the same waters as the growing numbers of Americans who were engaging in whaling and sealing in the Pacific and off the southern coast of Australia at the beginning of the nineteenth century. The first whaler to visit Australia was the *William and Ann*, a British vessel commanded by Captain Eber Bunker of Nantucket, which arrived in 1791. The first American vessel to call at Sydney was the *Philadelphia* in 1792, and American traders and whalers fishing in the Pacific visited that port regularly from then on. Bay whaling was initiated in Tasmanian waters in 1803 by Australian colonists. The typical deep-sea whale ship of the period was blunt in the bows with a cut-off square stern. She carried six or seven boats, five of which were slung over the bulwarks on heavy davits. The decks of the whale ships were fitted with brick tryworks, comprising large cauldrons with furnaces below. Two American vessels, the *Charles* and the *Union*, began sealing in Bass Strait in 1803. These and later visitors gave some concern to Governor King and his successors about the maintenance of adequate control over the territory by the British government.

The *Britannia*, belonging to the well-known British firm of Samuel Enderby and Company, was the first whaler known to have gone down in Australian waters. This vessel was wrecked on one of the reefs to the north of Lord Howe Island in 1806.[18] Another whaler from Britain, the *Echo*, en route to New Zealand, was wrecked in Torres Strait in 1820. The first Hobart-built whaler to be wrecked was the *George*, which went down off Lord Howe Island in 1830. Thereafter, during the 1830s and 1840s, a large number of whalers—French, British, American and Colonial—were lost on the east, south and west coasts of Australia. The American whaler *Thomas Nye*, built in 1851, was later used in general trading off the Australian coast as the *Day Dawn*. Later the vessel sank at Garden Island near Fremantle. The hull has been examined on the seabed by Western Australian Museum staff.

Another American whaler, the 314-ton barque *Charles W. Morgan*, has been preserved at Mystic Seaport, Connecticut. This vessel was built at New Bedford in 1841 and fished on some occasions in the Pacific. The *Charles W. Morgan* may be regarded in some respects as typical of the American whalers of the 1830s and 1840s. However, it would be wrong to regard one preserved ship as providing all the answers to questions about her type and period. Regular renovations must be carried out on any museum ship, and on each occasion alterations will be made according to interpretation by particular shipwrights. The vessel has not remained identical since the 1840s, as in some respects a shipwreck does. And it is likely that the shipwrecks of British, French and colonial whalers, and those of the local bay whalers, will throw further light on aspects of the industry in Australian waters.

One such whaler wreck has recently been found on a coral atoll off the north-west coast of Australia. The site has not yet been positively identified but current indications are that it is an English or European vessel, wrecked around 1810. An interesting feature of the site is the presence of at least seven iron cannon.

Vessels which could be termed general traders were operated by private merchants to carry passengers from Britain and Ireland to Australia, and some of these ships were wrecked.[19] Excavation of well-preserved examples would provide information about the sorts of items the passengers carried with them. The worst shipping disaster of this period occurred with the wreck of the 710-ton ship *Cataraqui*, carrying 46 crew and 369 immigrants, principally from Bedfordshire, Staffordshire, Yorkshire and Nottinghamshire. The *Cataraqui* struck King Island in Bass Strait during a strong gale on 4 August 1845, and 406 people perished. The ship struck at about 4.30 a.m. and with the sea breaking right over her a scene of the utmost confusion ensued.[20] Half an hour later the ship tipped over on her port side, and the boats were carried away. At about midday the ship parted amidship. The several survivors got ashore by clinging to floating wreckage. Vessels bringing passengers to newly established colonies were particularly prone to accident. The ship *Solway* reached Rosetta Bay in South Australia in 1837 with 52 German immigrants for the

Barrels of provisions lie stacked on the wreck of the *William Salthouse* (1841).
(Photo: Mark Staniforth)

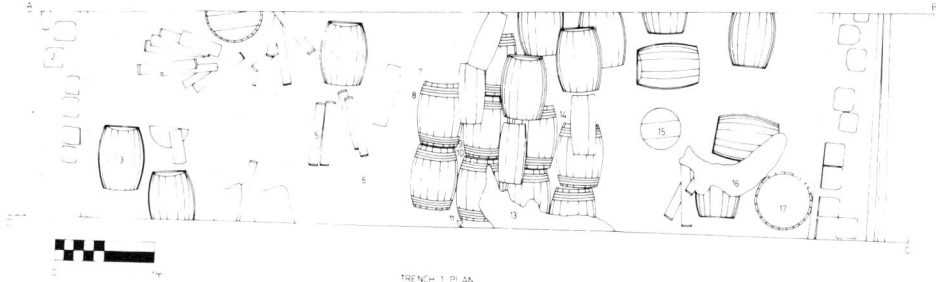

A section of the *William Salthouse* site plan shows the disposition of the barrels. (Drawing: Victoria Archaeological Survey)

Marks on the barrel lids from the *William Salthouse* describe the contents and indicate place and date of packing. (Drawing: Geoff Hewitt)

South Australian Company's service, and was driven ashore in a gale. The ship *Rockingham*, chartered by the Western Australia Company, arrived in Cockburn Sound in 1830 with around 200 passengers, and was driven ashore.

General traders formed the vast bulk of shipping arrivals during the first half of the nineteenth century.[21] Vessels from Britain brought manufactured goods and supplies, and either left in ballast for Asian waters or returned to Britain with produce from the whalers, sealers and farmers. Those from Asia brought a variety of goods including grain, tea and sugar, salted provisions, cottons, cloth, cattle and the infamous rum from India, and left with cargoes which sometimes included sandalwood and other timber, coal and bêche-de-mer. The many wrecks of these general merchant vessels contain representative cargoes from almost every year between 1800 and 1850.

An example is the 251-ton brig *William Salthouse*, wrecked near the new Port Phillip Bay settlement in 1841, with a cargo of provisions from Canada including salted pork, beef and fish, barrels of flour, and crates of wine and champagne, as well as nails, timber and iron bars.[22] Another is the 518-ton barque *Grecian*, lost at the entrance to Port Adelaide in 1850. The *Grecian's* general cargo from England was part salvaged in 1872.

The East Indiamen—those passers-by which were sometimes cast up on the Australian coast during the seventeenth and eighteenth centuries—grew in number during the early nineteenth century. Some followed the new route, continuing east past the southern coast of Australia, and then, sometimes, calling briefly at Sydney on their way north to China. But others continued to follow the route laid down by the Dutch in the early seventeenth century, and several were wrecked on the Australian coast in the early years of the nineteenth century.

LATER DEVELOPMENT: 1850 TO THE PRESENT

The heaviest concentration of shipwrecks during the later nineteenth century was along the New South Wales coast. Jack Loney, in his *Australian Shipwrecks*, lists some 950 shipwrecks off New South Wales, 575 off Queensland, 375 off Tasmania and in Bass Strait, 350 off Victoria, 275 off Western Australia, 175 off South Australia and 25 off the Northern Territory.[23] These figures cannot be regarded as comprehensive, but they do, in the absence of detailed statistics, provide a rough guide.

The occurrence of shipwrecks during this period is not solely a reflection of the shipping tonnage visiting each State. There were more losses off the coasts of Queensland and Tasmania (when the dangerous Great Barrier Reef and Bass Strait are taken into account) than off the Victorian coast. Western Australia with its very long coastline claimed more casualties than the more populous South Australia. The numbers of casualties appear to be influenced by the volume of shipping, the coastline, the climate, and the particular trades and industries among other things. With the new interest in maritime archaeology in each State there is now a large body of data on site locations, and this needs analysis to discover why the casualties occurred.

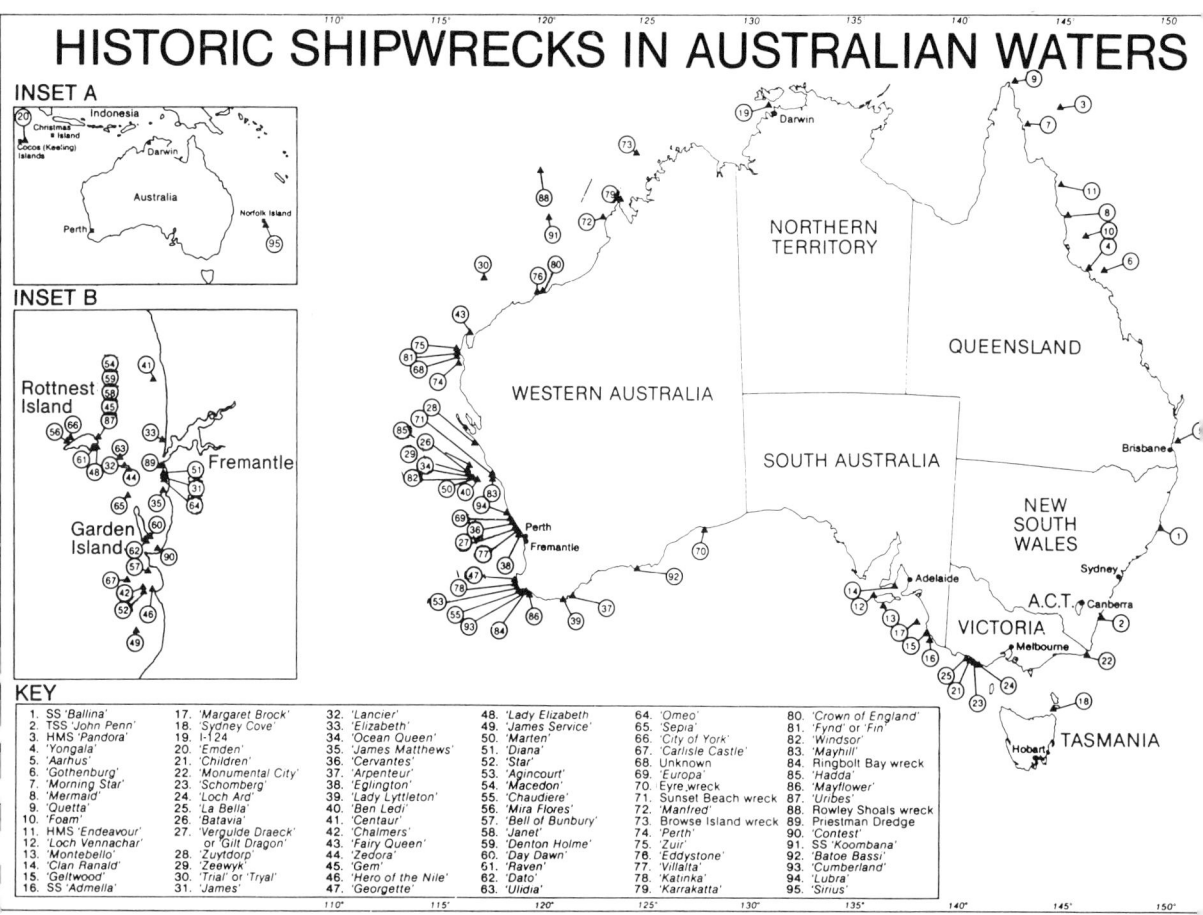

Sites gazetted under the Commonwealth Historic Shipwrecks Act to the end of 1984.
(Reproduced courtesy Department of Arts, Heritage and Environment)

During the latter half of the nineteenth century a number of rapid and far-reaching changes occurred in the technology of sail. These changes were so important that some maritime historians have seen the period as the golden age of sail. The wooden clippers (the term 'clipper' was derived from the word 'clip', meaning 'pace' as in 'to go at a good clip')[24] of the 1840s and 1850s were succeeded by the composite wood and iron ships of the 1860s and the large iron clippers of the 1870s and 1880s. They in turn led to steel ships—sturdier and more economical than the dainty clipper ships. They could carry six times the tonnage with fewer than twice as many crewmen. These developments were not pioneered in Australia, but the trends soon permeated Australia's overseas shipping, and the long route to Australia provided some of the stimulus for technological improvement.

Among the timber built clippers lost were the 2,284-ton Black Ball Line ship *Schomberg* (commanded by the notorious speedster, Captain 'Bully' Forbes), wrecked off Victoria in 1855; and the 2,028-ton *Lightning*, burnt at Geelong, Victoria in 1869. The *Schomberg*, an extreme clipper, was the largest ship of the 1850s built in the United Kingdom—equal in size to the largest American clipper. The *Lightning* was a remarkable vessel. The first clipper ever designed and built in America for an English firm, she was made for James Baines and Company of Liverpool, who contracted with Donald McKay to build several first-class ships for their Australian Black Ball Line.[25] *Lightning*'s sailing feat of 436 nautical miles (some 700 kilometres) in 24 hours, under Captain Forbes in 1854, makes her one of the fastest ever ships under sail. Not until 30 years later did an ocean steamship exceed her day's run.

The clipper ship *Schomberg* leaves England for Australia.

Some composite ships (iron ribs and wooden planking provided strength and extra space for cargo) were wrecked off the Australian coast. An example is the 1,106-ton ship *Tubal Caine*, which sank after collision off the Victorian coast in 1862. The *Cutty Sark*, preserved at Greenwich, England, is a survivor of this form of construction, and indeed the sole survivor of the clipper era. She combined speed with a good carrying capacity and great endurance. In 1884 she sailed home to London from Australia in 80 days. Between 1885 and 1894 this clipper lugged some 46,000 bales of wool, weighing a grand total of 18.6 million pounds (8.4 million kilograms), from Australia to ports in England and Belgium.[26] The last composite vessel built in Great Britain, in 1876, appears to have been the barque *Helena Mena*, a regular in the London to Fremantle trade during the 1880s.[27]

Hundreds of iron clippers were wrecked. Notable examples are two Loch Line vessels: the 1,693-ton *Loch Ard*, wrecked off the Victorian coast in 1878; and the 1,552-ton *Loch Vennachar*, wrecked on Kangaroo Island, South Australia in 1905. The wrecks of iron vessels have to date received much less attention from maritime archaeologists than their wooden counterparts. This is because maritime archaeologists have concentrated their interests on the period prior to the introduction of iron and steel ships.

The iron and steel shipwrecks lying in Australian waters demand urgent attention from our cultural resource managers. It is probable that in terms of volume they form the largest part of our underwater heritage: such vessels have constituted most of the tonnage plying our coasts since about the turn of the century, and the individual shipwrecks are much bigger on average than the wooden shipwrecks. In addition, in most Australian marine biophysical environments an iron ship will remain intact for much longer than a wooden ship, and will provide correspondingly greater protection for its contents, such as fittings, crew's possessions and cargo.

Unfortunately, this means that the iron wrecks are more conspicuous on the seabed, and those not fully protected by legislation and surveillance have become the focus of attention of souvenir-hunting divers seeking attractive complete artefacts and scrap metal. The problems of archaeological management are exacerbated because a significant proportion of the general public does not yet recognise these later and more complete structures as a genuine cultural resource.

This problem of lack of appreciation is not restricted to Australia. In 1983 I went to New Zealand to give evidence as an archaeologist in a test case involving the legislation protecting historic shipwrecks. A diver was then convicted of wilfully modifying an archaeological site, the wreck of the iron steamship *Taupo* which sank in 1881. Later a High Court judge quashed the conviction, stating that the defendant could not be expected to have known that the wreck was an archaeological site, and commenting that it was hardly in the class of the Waitangi Treaty house (in the crucial Waitangi Treaty Maori chiefs ceded land to Great Britain in 1840) as an historic place.

In all Australian States there are iron shipwrecks, relatively intact, which are visibly deteriorating annually because of such plundering. Iron and steel wrecks do present problems for archaeologists. Iron on the seabed has its form obscured by corrosion, concretions and marine life, making survey work difficult. Moreover, the iron, unlike wood, becomes brittle on the seabed, making any human activity on an iron site frustratingly destructive. And iron objects raised from a marine environment still present horrific problems for conservators. Nevertheless, these iron structures are clearly under great threat with the growing popularity of skin-diving, and they require particular attention by site managers.

As Australia grew, so the vessels employed in her trade grew larger. The wool industry prospered, and shipping received a massive stimulus from the discovery of gold and other minerals. Then later in the century the emerging exports of wheat, coal and timber added to the sailing vessel tonnage required. With the greater tonnage of sailing vessels being employed there were more shipwrecks, representing the diversity of industries in the cargoes they contained.

The ship *Duke of Bedford*, loaded with wool for London, disappeared after clearing Port Phillip Heads, Victoria, in 1852. The Blackwaller *Madagascar*, with 63,390 ounces of gold dust on board, disappeared after leaving Hobson's Bay, Victoria for England in 1853. The iron ship *Star of Greece* was carrying 16,000 bags of wheat bound for England when she was blown ashore near Port Willunga, South Australia, in 1888. The American barque *Loda*, from Melbourne to Shanghai with a cargo of coal, was destroyed by fire off the Queensland coast in 1886. And the Norwegian barque *Arcadia*, loaded with West Australian timber, was blown ashore at Hamelin Bay in 1900. Many hundreds of other examples could be cited of ships lost while carrying away the produce of Australia.

Nor should our most important import—people—be ignored during this period. A speaker at a recent conference of Australian archivists said he expected:

> that in the next several decades, Australian historians will get around to researching and writing the history of immigration over the last 200 years. I find it astonishing that this country has not produced an Australian Oscar Handlin or Marcus Lee Hansen to write the equivalent of the great histories of immigration into the North American continent. This must be one of the most salient themes of Australian history but is yet unarticulated in any satisfactory sense. It is a marvellous theme for any ambitious historian.[28]

So far our archaeologists have only been interested in the prehistory of immigration. However, a large material source exists for later studies—in the wrecks of passenger carrying ships. The discovery of gold brought a spectacular increase in the number of people who came to Australia in ships. Australia's population doubled in five years as clipper ships serviced the immigration boom. The passenger trade was carried mainly in ships owned by Liverpool firms, although most of the vessels had been built in America. These sailing vessels were not displaced by steam until the 1880s.

The transition from sail to steam was a technological development of great importance. Spectacular advances were made in steam technology during the nineteenth century, the merchant steamer developing from a small wooden inland waterways vessel propelled by paddle wheels and a simple engine, to a large steel vessel propelled by twin screws and driven by triple expansion engines of high efficiency and great reliability. A major advance came in the 1860s with the perfection of the compound engine. The second cylinder resulted in a considerable increase in the amount of power to be derived from a given unit of steam. Boiler pressures increased while coal consumption was cut down. However, it was the perfection of the triple expansion engine which led to the final demise of the sailing vessel. The necessary cheap steel was not available for the boilers of triple expansion engines until the 1870s, but in the 1880s the quality increased while the price decreased, and by the early 1890s pressures of 200 pounds per square inch (14 kilograms per square centimetre) were being achieved in steamship boilers. The triple expansion engine greatly reduced fuel consumption, reduced the weight of the engine and the amount of coal which had to be carried (increasing carrying capacity), and permitted a speed which allowed steamers, which were growing in size much more rapidly than sail, to make on average three times as many voyages as a sailing ship within a given time.

As the steam engine improved, sail was displaced first from the shorter trade routes, and then from the long-distance deep-sea trades. The establishment of coaling stations, initially supplied by sailing vessels, enabled steamships to reach almost anywhere in the world, including Australia. By the 1890s steamers were encroaching on the Australian wool trade, just as they had earlier done in the opposite direction with mails, passengers and manufactured goods.

The traditional steam engine itself has now all but disappeared from the sea lanes, replaced by more efficient power systems. A good deal was written about the development of steam during the period of its ascendancy, but it is certain that in the future historians will want to know more, and when they do they will have to look to the vessels lying preserved on the seabed. Large blocks of iron will survive much longer on the seabed than will iron plates (because of their concentrated mass), so steam engines will outlast iron hulls. Steam engines lying on the seabed thus provide the opportunity for examination of intricate engineering details.

A representative collection of steamships has been partly preserved by the sea on the seabed. Through examination of these sites we can go back to the very early phases of steam power.

The 250-ton paddlewheel steamer *Ceres* appears to have been the first steamer lost in Australian waters. She was wrecked north of Port Jackson in 1836, but her engines were later raised, and the wooden hull taken ashore. The 297-ton *Clonmel*, a wooden-hulled schooner-rigged paddle steamer built in 1836, was one of the first steamers brought to Australia for coastal trade, and was claimed to be the first seagoing steam vessel in the Pacific.[29] She was wrecked on the Victorian coast in 1841, and her boiler is still exposed at low tide near the main Port Albert Channel. The 108-ton wooden paddle steamer *Phoenix*, initially fitted out with machinery from the *Sophia Jane* (Australia's first steamship), was wrecked on the Clarence River bar, New South Wales, in 1852. The 768-ton American steamship *Monumental City*, the first steamer to cross the Pacific under screw propulsion, struck a reef near Mallacoota on Victoria's east coast in 1853, and sank with the loss of 37 lives. The launching of the screw steamship *Archimedes* in 1840 had heralded the end of the

Passengers at Fremantle. (Photo: Battye Library 5323B/46)

paddle steamer as an ocean carrier. The first decade of screw steamers saw the introduction of a variety of engines designed to better suit the new mode of propulsion. The wreck of the *Monumental City* provides an early example of the use of direct acting oscillating steam engines for such vessels.

The number and variety of steamship wrecks increased towards the end of the nineteenth century as steamship owners wrested the coastal and overseas trades from sailing ships. Sometimes steamships were wrecked because their masters and owners had inadequate knowledge of the requirements and hazards of the routes. Other owners did things on the cheap, purchasing worn out vessels with obsolete machinery and placing them on routes which did not yet have sufficient cargo tonnage to warrant the additional cost of a steamship. Old vessels introduced to the rigors of the Australian coast frequently lasted no more than a year or two. Examples of this were the 66-ton SS *Xantho*, which was 23 years old when introduced to the coast in 1872, and lasted six months before sinking; and the 48-ton SS *Sunbeam*, a 29-year-old vessel which lasted a little over a year after her introduction in 1890. These two vessels sank without loss of life but other steamers were lost in less fortunate circumstances. The 3,481-ton British India Steam Navigation Company's screw steamer *Quetta*, for example, was wrecked in Torres Strait in 1890 with a loss of 133 lives.

During the transitions from sail to steam and from wood to iron, some curious developments took place in naval vessels. The Victorian Government, interested in the activities of the American commerce raider *Shenandoah*, and imagining a Russian threat to Port Phillip Bay, requested that a suitably armed man-of-war be stationed there.[30] The British Government then constructed a 68-metre-long iron and steel armour-plated monitor or turret ship which was named *Cerberus*. After a difficult voyage out from Britain via the Suez Canal in 1871, the *Cerberus* underwent trials, and then spent most of the next 50 years quietly rocking on her moorings off Williamstown. She was sold to a Melbourne salvage company in 1924 and later scuttled in shallow water to form a breakwater. A *Cerberus* Preservation Trust was formed in 1970 with the aim of raising and restoring the hulk, but nothing has eventuated.

The study of shipwrecks of the twentieth century may seem to some people too modern a task for archaeologists. Many archaeologists, perhaps concerned about its intellectual validity, tread warily in post-medieval studies. The journal *Post Medieval Archaeology* has the year 1750 as its terminal date of interest, and the *International Journal of Nautical Archaeology* in its early issues shows signs of a similar wariness of the more recent past.[31] Even today, many archaeologists would not contemplate involvement in a period for which there are still living representatives, despite the rapid growth of ethnoarchaeology. But any significant aspect of the past which is not adequately covered by documentary evidence or which might be open to a different interpretation through examination of material remains should be fair game for the archaeologist. Even a site only several years old may have the potential to answer questions about human behaviour through the application of archaeological method

The study of the 'evolution' of the pearling lugger in Australia is one example of a useful recent cultural resource. Pearls and pearl shells have been gathered along the north coast from the 1850s, and the vessels known as 'luggers' have gone through a number of changes with expansion of the industry and the availability of more advanced technology, since the 1870s. Today there are less than a dozen pearling luggers operating in the industry in the north, and these are being phased out as new approaches bring fundamental changes in vessel shape, size, and construction materials. Most of the surviving pearling luggers were built soon after World War II, and are the culmination of a century of design development. The available documentary record fails to give an adequate account of these changes, so it is through discussion with people involved in the industry, and a study of the vessels still afloat, those preserved as museum displays, the old hulls lying in the mangroves and the wrecks lying on the seabed, that the story can be revealed. An archaeological study of a maritime community such as the pearl shellers of Australia needs to involve both underwater and terrestrial sites. Other aspects of the fishing industry could be studied using a similar approach.

The pearling lugger *EWS* on a Museum inspection expedition to the Kimberley tide regions. (Photo: Pat Baker)

A postcard representation of the Aberdeen White Star Line's 11,400-ton TSS *Pericles* going down. (Photo: Western Australian Museum)

By the early twentieth century passenger liners had been equipped with comforts such as electric lighting and refrigeration, and had reached massive proportions— ships of 18,000 tons and more bringing passengers to and around Australia in a state of luxury previously not even dreamed of. Some of these vessels sank in deep water and are well enough preserved to give investigating archaeologists numerous insights to sea travel at that time. Many have been found by divers. These include the Adelaide Steamship Company's 3,663-ton *Yongala*, built in 1903 and lost off the Queensland coast in 1911; the British India Steam Navigation Company's 5,156-ton *Satara*, built in 1901 and lost off the New South Wales coast in 1910; and the Aberdeen White Star Line's 11,400-ton *Pericles*, lost off Cape Leeuwin in 1910. The *Yongala*, in particular, is in an excellent state of preservation and, if souvenir hunters can be dissuaded from interfering with the wreck, it has the potential to yield a great deal of information about coastal travelling at the beginning of this century.

The waters around Australia have been disturbed by relatively few shots fired in anger, but there are nevertheless some defeated naval vessels lying offshore.

EMDEN.

The battered German light cruiser *Emden* lies in the surf at Cocos Keeling in 1914, destroyed by HMAS *Sydney*. (Photo: Western Australian Museum)

Examples are the Japanese submarine *I.124*, sunk by the Royal Australian Navy off Darwin during World War II, and HMAS *Sydney* sunk off Shark Bay. The 6,380-ton cruiser *Sydney* was sunk during an engagement with the German raider *Kormoran* on 19 November 1941. It was Australia's greatest and saddest naval loss. Only the examination of the wreck itself (and it has not yet been located) can end the speculation about how the *Kormoran* managed to sink this powerful ship. Both of these sites lie in relatively deep water, precluding archaeological excavation in the near future.

An anthropologist recently used official enquiry records relating to an ocean-going oil rig disaster—that of the *Ocean Ranger* in the early 1980s—to look at a variety of aspects of human behaviour.[32] Observing that many of the crew were ex-farmers rather than seamen, and that the captain was subordinate to drilling experts on board, he went on to present the hypothesis that there are basic differences between the personality traits of seamen and landsmen. The landsman tames his environment whereas the seaman takes a more passive approach, riding with it. Thus the landsman is unsuitable as a commander of a vessel at sea. Other related questions could be explored by anthropological or maritime archaeological method. These sorts of questions relate not to the age of a shipwreck but to the circumstances of the loss and the condition of the site.

Chapter 4   REFERENCES

1. Crowley, 1974, pp. 1-6.
2. Martin, 1978, p. 143.
3. Blainey, 1966, p. 44.
4. Historical Records of New South Wales, II, p. 706.
5. Historical Records of New South Wales, II, p. 761.
6. Historical Records of New South Wales, II, p. 757.
7. Hough, 1972, p. 106.
8. One man was, however, killed at Tofua before the boat voyage began.
9. Rawson, 1963, p. 61.
10. Taken from Lyon, in Henderson, Lyon and MacLeod, 1983, pp. 28-35.
11. Blainey, 1966, p. 61.
12. Historical Records of New South Wales, III, p. 758.
13. Bateson, 1972, p. 29.
14. Calder Papers, p. 301.
15. Bateson, 1972.
16. Broeze, 1975, p. 583.
17. Historical Records of Australia, I, p. 254.
18. Bateson, 1972, p. 43.
19. Broeze, 1982, pp. 235-253.
20. Bateson, 1972, p. 187.
21. Broeze, 1978, p. 194.
22. Staniforth and Vickery, 1984.
23. Loney, Vol. 2, 1980, and Loney, Vol. 3, 1982.
24. Whipple, 1980, p. 21.
25. Loney, Vol. 2, 1980, p. 217.
26. Whipple, 1980, p. 150.
27. MacGregor, 1973, p. 162.
28. Richards, 1984, p. 23.
29. *Sydney Herald*, 3/12/1840.
30. Maritime Archaeology Association of Victoria, 1983, p. 3.
31. See the editorial, *Post Medieval Archaeology*, 1967, and the *International Journal of Nautical Archaeology*, 1974.
32. Wright, Guy, unpublished seminar presented to Anthropology Department, University of Western Australia, 1984.

5 Finders and the Law

In Australia the governments of today generally take the attitude that maritime archaeology is a public responsibility. There is no such thing as 'private archaeology'—it is a proper concern for everyone. There is a growing recognition that archaeologists and other concerned citizens of today who fail to act quickly and positively must share the blame for any loss of our heritage equally with those who are actively, even if unwittingly, destroying the record of the past.[1] Clearly, legislation is necessary to protect the resource. For legislation to be successful it must be seen by the general community not only to be worthwhile, but also to be fair. The aim of course is to protect archaeological sites for the public, but in doing so any legislation cannot afford to ignore common rights of individuals. The rights of the shipwreck finder are often debated in diving circles, so in this chapter about the law I have tried to clarify the finder's position in Australia.

There are two types of shipwreck finder: the diver who deliberately sets out to locate previously unknown sites, and the one who stumbles, quite by accident, upon a site. The former is generally fully aware of his rights and obligations under the law, and of any likely archaeological or historical significance attached to the site. But the unintentional discoverer is frequently confused. Is he the first discoverer, or is the site known to others? Is it historically or archaeologically significant? Does it contain items worth salvaging? Is it covered by legislation of some kind? Should he remove any items? Should he report the find to a government authority?

I advise the diver to follow these steps. Take note of the nature of the wreck. Is it constructed of ferrous metal or wood? The exposed planks and frames of a wooden shipwreck usually disintegrate after some years, leaving metal fastenings lying on the seabed. How large is the wreck? Are there recognisable cargo items which might give clues to the age, function and identity of the ship? What is the nature of the seabed and the depth of water? Is there a plaque, wreck buoy or other indication that the site is known to the government? Take note of the location of the site including proximity to reefs or islands. Record compass bearings and photograph transits to known points on land and, if you can, mark the latitude and longitude on a chart of the area.

Historic shipwrecks in Australian waters are protected under heritage legislation, and other shipwrecks are covered by the Navigation Act, so it is necessary to establish the identity and status of a shipwreck before altering it in any way. You can check in your local library for books about the historic shipwrecks in your area, or telephone one of the government departments dealing with maritime archaeology. That should indicate whether or not you have found a previously undiscovered site, and whether it is protected under heritage legislation. If you have found a new historic shipwreck you should tell the relevant government department (the Minister for Arts, Heritage and Environment if the site lies in Commonwealth waters, or one of the State departments if in State waters), and complete a shipwreck finder's form for that department.

The original Historic Shipwrecks Act, 1976, did not automatically fully cover all shipwreck sites in the waters adjacent to the coast. The Act required the Minister to formally declare each wreck he considered should be protected. This placed some onus upon concerned citizens in each State to speak up if they felt that a significant site needed protection. A 1985 amendment to the Act, however, enables the Minister to declare all remains of ships (whether or not the existence and location of the remains are known) situated in Australian waters which are at least 75 years old to be historic shipwrecks. Other criteria currently suggested by the Department of Arts, Heritage and Environment (the Commonwealth department responsible for administering the Act) for deciding whether particular wrecks should be declared historic are as follows:

(a) a wreck significant in the discovery, early exploration, settlement or early development of parts of Australia;
(b) relevance of a wreck to the opening up or development of parts of Australia;
(c) relevance of a wreck to a particular person or event of historical importance;
(d) the wreck is a possible source of relics of historical or cultural significance;
(e) the wreck is representative of a particular maritime design or development;
(f) naval wrecks, other than those deliberately scrapped or sunk and having no particular historical or emotional interest;
(g) wrecks of an outstanding recreational or educational potential.[2]

A maritime archaeologist will inspect and assess the site, and make recommendations as to its significance. You the finder may then be eligible for a reward. If you have a continuing interest in the site you may be able to join government maritime archaeologists in their activities, but that requires a permit.

If the shipwreck is not of any historical or archaeological significance and you wish to alter it then you should seek the advice of the Receiver of Wreck or consult the Navigation Act before recovering relics.

The difficulties of finding historic shipwrecks and raising valuable artifacts from them are not the most significant problems for the underwater cultural heritage, even if they are the most obvious.[3] Vast numbers of underwater sites have already been located and every week new finds are reported. The biggest problem is that of intentional human interference with sites after they have been found. To overcome this problem a country must enact protective legislation and implement an accompanying archaeological site management programme.

Any government which passes legislation and goes no further will fail in the task of protecting its underwater cultural heritage. Sports divers and the general community will only maintain a sympathy for such legislation, inevitably restricting as it is, if they are regularly given information about progress, by means of public lectures, publications, media releases and displays. And this requires State support for an active management and research programme.

For the individual archaeologist employed in government service legislation means that he can go ahead and open up a particular site for excavation when necessary without the fear that mid-way through the work he will find himself fending off uninvited guests: a team of salvage operators with equipment to raise commercially valuable items quickly and no desire to be distracted by the constraints of archaeology. Effective legislation also means for the archaeologist that he can leave the vast majority of sites untouched as a cultural resource for future generations to appreciate, and that he can give those sites to which he devotes his attention all the effort that they merit, rather than moving quickly from site to site, more with a view to saving precious objects from imminent destruction than with considered archaeological aims. Of course, legislation can also place a heavy burden of responsibility upon the archaeologist. In a country with a large number of sites, and a lengthy coastline, blanket protection might mean that he is put in the position of spending all his time inspecting and managing sites, leaving no time for more intensive investigations on any particular site.

In Australia, interest in maritime archaeology began in 1963 with the finding of two Dutch East Indiamen, the *Vergulde Draeck* and the *Batavia*.

Over the years the story of the loss of the *Vergulde Draeck* had grown into something of a legend on the west coast of Australia. It is unlikely that any of the original English settlers in the colony, established in 1829, knew of the wreck. But interest was sparked with the publication in 1859 of Major's *Early Voyages to Terra Australis*, a collection of translations of original documents in the archives at the Hague.[4] Other scholars during the latter half of the nineteenth century produced a number of books dealing with the discovery of Australia. These scholars laid particular emphasis on the Dutch voyages to Western Australia some 150 years before Captain Cook first examined the east coast. They also mentioned in passing that the *Vergulde Draeck* had been carrying, as well as a rich cargo of merchandise, eight chests of silver coinage amounting to 78,600 guilders, and that this treasure had been lost.

The treasure hunting started in earnest in 1931 after a boy found a chest of coins, probably from the *Vergulde Draeck*, on the coast near the mouth of the Moore River, the area where the Dutch search vessel *Waeckende Boey* had found wreck material in 1658. About 35 coins were found, consisting of ducatons and Japanese Mameita—and Cho-Gins, which indicated that they were a private collection belonging perhaps to an officer or merchant who had come ashore from the wreck, because the *Vergulde Draeck* carried reals as the official specie. Since that time a series of unsuccessful expeditions consisting of diviners, amateur historians and surveyors, and treasure seekers generally have combed the surrounding countryside, and many kilometres inland, seeking Dutch campsites, circles of stones and treasure.

In 1964, after most of the *Vergulde Draeck's* treasure had been found and removed from the wreck site, a syndicate nevertheless began digging for eight boxes of treasure on the mainland, at a place called Dynamite Bay. After 8 weeks of digging using a drag-line excavator, the top of the pit was 30 metres in diameter, and when funds ran out the barren hole was over 20 metres deep, far beyond the digging capacity of any Dutch castaway of the seventeenth century.

Apart from the treasure trove aspect, the discovery of the *Vergulde Draeck* was an event of considerable interest for maritime archaeologists, being the first modern underwater discovery of an outward-bound seventeenth century Dutch East Indiaman. Events moved quickly towards the enactment of protective legislation and the setting up of preliminary facilities within a State Government department to carry out maritime archaeology.

In Perth, the capital city of Western Australia, the discovery of the *Vergulde Draeck* was regarded as a sensation.[5] A newspaper immediately financed an expedition to the site, and soon a cannon, an anchor, numbers of stoneware jugs, ivory tusks and clay bricks, a sounding lead, a whetstone, sailmakers' scissors and silver coins had been salvaged. This material tended to confirm the identity of the wreck.

The finders of the wreck consulted local historians about the background of the ship and the wreck was declared to the Receiver of Wrecks, as required by the Commonwealth Government's Navigation Act of 1912. The provisions of the Navigation Act dealing with wrecks and salvage were derived largely from the British Government's Merchant Shipping Act of 1894, and the men who drafted these documents made no provisions for the protection of wrecks forming a part of a nation's cultural heritage. They were concerned with modern wrecks and could not have been expected to anticipate the results of advances in diving technology of the mid-twentieth century or the need for protection of archaeological sites underwater. The law in fact encouraged the destruction of historic shipwrecks, with its provisions relating to salvage and the requirement, where the ownership of a wreck could not be established within a year, that the Receiver of Wreck should dispose of the relics by public auction. Once removed from the water, many of the more delicate objects from historic shipwrecks would not last more than a day without appropriate conservation treatment.

If the legislative framework did not auger well for the future of the *Vergulde Draeck* as a maritime archaeological site, then the institutional situation in Western Australia was no more encouraging. The Western Australian Museum was the only organisation equipped to carry out extensive fieldwork on aspects of what might be termed the 'National Heritage'. This institution was essentially a natural sciences museum, but it had recently developed interests in Aboriginal studies, including prehistoric archaeology. It had not developed studies in Australian history or historical archaeology and had no capacity to respond immediately to the requirement to place staff and equipment in the field, so the wreck was not initially viewed by the Museum as any of its concern.

In this situation commercial salvage operators and sports divers could not be certain as to how the law applied to this wreck, that was so old, and apparently without any living owner. But divers interested in the historical aspects of the material could see that protection and the involvement of a funded institution were necessary. During the winter months of 1963, treasure-hunting scuba divers from all over the State joined a frenzied free-for-all on the wreck site as more chests of silver were uncovered.

Another discovery took place in June 1963. The *Batavia*, earliest of the Dutch East Indiamen wrecked off Western Australia, was found lying in shallow water off Houtman's Abrolhos, a group of coral islands some 480 kilometres north of Perth. The chain of events was similar to what was happening on the *Vergulde Draeck*: a responsible, organised private expedition raised material and positively identified the site, but soon after these divers had returned to the mainland clandestine looting commenced, with destructive effect upon the wreck site.

In October 1963 the news reached Perth of indiscriminate blasting on the *Vergulde Draeck* wreck site. A member of the police diving squad confirmed that explosives had been used, and other divers said that tunnels in the limestone reef had been made unsafe to work in. It now became apparent that the wreck would soon be completely destroyed unless the government could be brought to formally recognise the historical value of the site. The finders' group donated to the Western Australian Museum the relics they had raised, together with their rights of claim to the wreck as discoverers. This was formalised by a Deed of Assignment, drawn up by the Crown Law Department, vesting the rights of these people as finders in the State, which thereby became the legal guardian of the wreck.

When the Western Australian Museum first began to explore the possibility of drawing up legislation to protect archaeologically important shipwrecks, the State's legislators knew of no other country with adequate legislation. The increasing mobility of divers in Mediterranean waters had prompted Cyprus and Greece to enact legislation specifically referring to underwater items, and France had an act dealing specifically with some aspects of its underwater cultural heritage.[6] Neither the United Kingdom nor the Netherlands protected their historic shipwrecks.

The Bill drafted by the West Australian State parliament in 1964 was mainly concerned with the protection of the four known Dutch seventeenth and eighteenth century East Indiamen, and the one English (at that time undiscovered) seventeenth century East Indiamen. But it also included other ships wrecked up to the year 1900. It made provision for the reporting by divers of new finds, and for their assessment by the Museum, and it included reward clauses for reporters of new wrecks. The finder of a wreck was entitled to a reward not exceeding £1,000 and, if the wreck carried specie, he could also be paid the market value of the gold or silver, or the coin or bullion itself might be transferred to that person. On the other hand, there were penalty clauses for those who did not report finds, and for anyone who interfered with an historic wreck.

The Museum was given the responsibility to recover, preserve and display the wrecks. The Bill did not provide for rewards to people who had found wrecks before the legislation, such as the *Vergulde Draeck* and the *Batavia*.

During its passage through the State Parliament the most significant criticisms were directed towards legal aspects. One Member:

> ... forecast that this legislation would be wrecked on the battlefield of the High Court, and he issued a warning ... to approach the Commonwealth to see if it would not be the better authority to take over the control of these wrecks under the Commonwealth Navigation Act.[7]

The West Australian Government made a very bold and far-sighted step when it passed that legislation in 1964 for the protection of historic shipwrecks. The general public had previously tended to regard shipwrecks simply as potential treasure troves. Now the looters would be discouraged and the Museum would be able to survey, excavate and preserve the relics on behalf of the newly concerned community. The State legislation was to serve an important function, not in the negative aspect of prosecution, for no successful prosecution has been made under the State Acts in Western Australia (or under the Commonwealth's Act for that matter), but in establishing a moral basis for the work of maritime archaeology. If the Museum could be seen to be getting on with the job (displays and publications enable the public to see the work which has been done), then most people would prefer to abide by the laws which supported them in this. Today there is a wide acceptance of the concept of an Australian cultural heritage, and of the need for legislation in this area.

The Museum Act Amendment Act of 1964 (the first shipwrecks Act) gave the Western Australian Museum very wide responsibilities, but there were problems of implementation. A large number of wrecks were known to lie off the coast, and several of the earliest sites were extremely isolated, difficult to approach and difficult to police. An appropriate programme of activities could be expected to involve extensive inspection, survey and excavation work in the field, coupled with conservation, registration, display and publication in the office and laboratory. But progress in all these areas was slow.

In the field, it was only after equipment and an experienced team were co-ordinated by diving professional staff that archaeological results could be achieved. Prior to the excavation of the *Vergulde Draeck* in 1972, the role of staff was essentially a watchkeeping one. In the absence of experienced diving maritime archaeologists to direct excavations, the basic policy was correct: it was better for the Museum to attempt to maintain a static situation than to destroy sites through its own haste. But the policy alienated a large part of the amateur diving fraternity. The Government, the Museum, and the diving public learned by experience that the enactment of legislation does not in isolation protect culturally important shipwrecks. It does establish an involvement by Government, while it is the responsibility of the Museum, through its maritime archaeologists, to then establish a programme of excavation or other appropriate site management which will gain the support of the general public.

The Museum Act in Western Australia was amended in 1969, clarifying the position of finders. Gold or silver coins found on a wreck could no longer be transferred to the finder. The new legislation had the effect of bringing to the surface many of the problems which had fermented over the previous 5 years. In January 1970 a diver was charged under the Act with having removed part of the wreck of the *Vergulde Draeck*. A month later the same diver was committed for trial on a charge relating to the sale of copies of coins. In the same month the police seized approximately 2,000 silver coins.

In July 1970 an expedition to the *Batavia* site, financed by the University of Western Australia, found evidence of recent blasting. This was the last straw. The Museum prepared a submission to the Government indicating that it was impossible for it to carry on—it either had to have adequate support to employ curatorial staff and develop conservation facilities, or it should give up.[8] Fortunately, the Government gave the Museum the necessary additional support, and this ushered in a decade of outstanding success for the Museum's maritime archaeological programme.

Curators of Maritime Archaeology and Materials Conservation were employed and these two departments soon recruited balanced teams with practical and scientific expertise.

When in July 1971 it was reported that the *Trial* wreck had been blasted, the Curator of Maritime Archaeology, Jeremy Green, was able to mount an expedition to survey that remote site. A photogrammetric survey was carried out and the Museum was able to show positive archaeological results.

The following year saw an extensive excavation of the *Vergulde Draeck* wreck. This had been the most accessible of the Dutch wrecks to souvenir hunters but it nevertheless yielded to archaeologists a wealth of relics in good condition, including some 8,000 silver coins. In the rapidly expanding conservation laboratory the curator, Dr Colin Pearson, was able to treat ceramics and glassware immediately, and could store the problem materials—wood and cast iron—in stable solutions pending the development of satisfactory forms of treatment.

With the *Vergulde Draeck* excavation almost complete, the Museum was free to devote most of its resources over the next four years to the wreck of the *Batavia*. The wide array of seventeenth century artefacts on this site, and the presence of substantial ship's structure, exceeded all expectations. The *Batavia* proved to be the most productive archaeological site of all known seventeenth century Dutch shipwrecks.

In 1973 further legislation was prepared in Western Australia, aimed at getting away from some of the concepts involved in the Navigation Act. The new legislation (the Maritime Archaeology Act, 1973) removed the emphasis on the act of wrecking—that is, being cast ashore—and introduced the concept of 'historic ship'; that is, any ship lost, wrecked, abandoned, or stranded before 1900, whether above or below the low water mark.

No charges have been laid under the 1973 Act. The Museum's stepped up fieldwork programme had convinced the majority of the diving fraternity, as well as the general public which visited the Museum's displays, that archaeological excavation was the only way to deal with historic wrecks under threat. More and more people

The Western Australian Maritime Museum, where the remains of the hull of the *Batavia* (1629) are housed. (Photo: Pat Baker)

accepted that it was better for the State to professionally excavate, conserve, research and display archaeological treasures in its museums than for individuals to hoard the items on their mantlepieces until they disintegrated through lack of treatment. Many divers now preferred to join with the Museum in its fieldwork so as to experience the exhilaration of seeing new finds—with no desire for personal ownership.

The excavations of the Dutch shipwrecks, and several post-settlement sites, yielded so many objects that both the conservation laboratory and the Fremantle Museum were filled to capacity. The Commissariat building, a large mid-nineteenth century stone warehouse situated near the sea in Fremantle's historic West End, was therefore acquired as a maritime museum and opened during Western Australia's sesquicentennial year celebrations in 1979.

The West Australian shipwreck legislation was finally given the test constitutionally in a High Court of Australia action in 1976, when a salvage diver from Western Australia challenged the validity of the Act. It was contended that provisions of the Act were invalid on the ground that they were inconsistent with the Commonwealth Navigation Act and were therefore inoperative under the provisions of Section 109 of the Constitution of the Commonwealth. In August 1977 the High Court declared certain key provisions of the Maritime Archaeology Act invalid on constitutional grounds: that it was beyond the powers of the State of Western Australia to legislate in respect of historic ships situated in the territorial seas of Australia; this was properly a matter for the Commonwealth Government.[9]

The High Court proceedings in 1976 had prompted the Australian Government to present a Bill for the protection of historic shipwrecks to Parliament, so that if the High Court decision went against the West Australian legislation, there would not be a 'free for all' situation. Ironically, the challenge to the State legislation led to the passing of the Historic Shipwrecks Act in December 1976, extending the protection of historic wreck sites to the Commonwealth level.

The greatest legal problem in terms of the protection of historic shipwrecks in Australia has been the constitutional issue. Why then was the legislation not Commonwealth to begin with in 1964? At that time the Governments' legal advisors, both State and Commonwealth, were of the opinion that the State Government could legislate in the area. Besides, the Commonwealth Government showed no interest in legislating to protect historic shipwrecks. The Dutch wrecks were considered very much a part of Western Australia's history: a vital collection of relics which illustrate in a unique and dramatic fashion those European contacts with Australia some 150 years before Captain Cook's arrival.

Soon after the passing of the 1964 Act the constitutional opinions had begun to change. In 1965 D. O'Connell's *International Law in Australia* was published, with a foreword by the Chief Justice of the High Court, Sir Garfield Barwick. O'Connell argued that the low water mark was regarded for most purposes during the colonial period as being the territorial limit, and that with Federation this position remained unaltered, while the Commonwealth assumed responsibility for legislating outside that boundary.[10]

Until O'Connell argued that the definition of the States' territories began at the low water mark it had been accepted that the States had jurisdiction over the seabed to the 3-mile limit. This was in effect the territorial limit pre-Federation, and one of the reasons for Federating had been the desire of the States to get governmental control extended beyond the 3-mile limit. Western Australia and Queensland were particularly concerned at the time of Federation to see Australian control over foreign exploitation of the pearl and bêche-de-mer resources in the northern waters.[11]

Prior to Federation there was a Council for Australasia and the British Parliament passed an act which empowered this Council to exercise control beyond the 3-mile limit, so there was no question as to the States' position.

Following O'Connell's publication, there was debate as to the real jurisdiction of the State. Particularly during the prime ministerships of Gorton and Whitlam, the Commonwealth and State governments were in conflict about the right to legislate for offshore oil and mineral exploitation. This culminated in the Commonwealth's Seas and Submerged Lands Act, 1973, which was subsequently tested and upheld in the High Court. Thus, viewed historically, there was a clarification of the limits of State and Commonwealth jurisdiction during the first 13 years of operation of the State's Acts. It was now clear the Commonwealth had control over the area from low water mark to the edge of the continental shelf, except in bays. An indentation was not to be regarded as a bay unless its area was as large as, or larger than, that of the semi-circle whose diameter was a line drawn across the mouth of the indentation.

Obviously, while there had still been hope that the States could retain the more extensive rights, the Western Australian Government would not jeopardise its position by asking the Commonwealth to legislate on historic wrecks. So the challengeable legislation of 1969 and 1973 was persevered with. Nevertheless, after negotiations between Canberra and the Dutch authorities on the one hand, and the State Government of Western Australia on the other, an agreement was signed late in 1972 whereby the Netherlands Government vested its rights of recovery of the Dutch East India Company wrecks (questionable though they might be) in the Commonwealth of Australia, which in turn delegated the authority to the Western Australian Museum. The modern Dutch Government became the direct heirs of the old East India Company in 1795, when it was nationalised, and so claimed owners' rights over the Dutch wrecks off the West Australian coast.[12]

In 1974 a conference of scholars, museum workers and historical archaeologists was called together in Canberra under the auspices of the Interim Committee on the National Estate (later replaced by the Australian Heritage Commission). The purpose of the conference was to consider the principles, problems and methods of preserving historic sites, including shipwrecks, in Australia. At the meeting it was moved by a West Australian delegate (the author) that since the provisions of the State legislation in Western Australia were bringing about good work the Commonwealth Government should be asked to consider similar legislation, so that all States would be able to protect the potential of their maritime archaeological sites. During the same year the Commonwealth Government, as part of an examination of proposals for offshore legislation in various fields, directed the Special Minister of State, as he then was, to assume responsibility for the control of historic wrecks around Australia.[13]

Faced with the problems of preparing legislation, the Commonwealth Government formed an inter-departmental committee. Thus, when Parliament was spurred into action in 1976 to circumvent any problems regarding site protection which might arise from the High Court challenge to the West Australian Act, the inter-departmental committee had already given years of planning to a Bill which had drawn upon the much-improved West Australian legislation of 1973. This enabled a relatively smooth administrative transition from State to Commonwealth legislation.

Australia now leads the world with its legislation and site management programmes for the underwater cultural heritage.

Some observations need to be made about the operation of the Historic Shipwrecks Act. When passed in 1976 the Act applied only to Australian Territories. The Commonwealth adopted a practice that the Act would not be applied to the waters of any State without that State's concurrence. The Act was proclaimed in Western Australia and Queensland in 1977, in New South Wales in 1979, in South Australia in 1980, and in Tasmania and Victoria in 1982. The relics recovered from the Dutch wrecks off the coast of Western Australia are protected under the separate agreement of 1972 between Australia and the Netherlands. The Australia-Netherlands Committee on Old Dutch Shipwrecks (ANCODS) determines the disposition and subsequent ownership of these relics. In practice, the bulk of the collection has remained in Western Australia, while sample collections have gone to Canberra and Holland. In 1983 a human skeleton and silver coins worth $29,000 were among property handed over to the Netherlands Government.

Sites lying above a State's high water mark or in a State's bays are not covered by the Historic Shipwrecks Act, but South Australia, Victoria and Western Australia have separate acts covering these residual areas.[14] The State Acts assume greater importance in States whose bays contain large numbers of shipwrecks, such as South Australia with Spencer Gulf and Gulf St Vincent, and Victoria with Port Phillip Bay. Currently, the Acts contain differences in the sorts of sites protected and the sort of protection given. It is likely that they will become more consistent in the future.

The Historic Shipwrecks Act provides for rewards to be given to people who first notify the Minister of the discovery of an historic shipwreck. The provision of rewards is intended as an incentive for people to come forward with details of wrecks so that the sites can be fully protected. The Commonwealth puts it this way:

> While many would argue citizens should comply with the law without the inducement of rewards, I think you will agree that in view of human nature this approach is unrealistic.[15]

The argument against rewards being given might go something like this. We have legislation. The historic shipwrecks belong to the State. It is demonstrably good legislation designed to protect the cultural heritage for the people of Australia. Legislators have not provided rewards as incentives for people to report finding dry-land sites. The shipwreck legislation provides penalties for non-disclosure of finds, and police forces exist to enforce the legislation. So if people do not comply they can be prosecuted. Divers do not therefore need to be given rewards for compliance.

Hopefully there will come a time when pride in our heritage will be sufficient inducement for people to report these sites—indeed this is already the normal situation in terms of the attitude of finders. But for some finders today a temptation would be felt. Policing of this aspect of the legislation is difficult, and if rewards were abolished now the necessary increase in surveillance would very likely bring increased overall costs, together with a confrontationist situation between some divers and those implementing the legislation.

There are only four known bullion-carrying shipwrecks on the Australian coast—the *Batavia* (1629), the *Vergulde Draeck* (1656), the *Zuytdorp* (1712) and the *Rapid* (1811). Salvage of these sites by their finders might have been very profitable. But of all the other historic shipwrecks on the Australian coast few would return much profit to a salvager. So, from the finder's point of view, a reasonable reward for notification, and the satisfaction of involvement with the archaeological aspect of an historic shipwreck, is the preferable course to concealment or looting with the threat of the disgrace of prosecution. No guidelines are given in the Commonwealth's legislation to assist in determining the size of the reward, but the Minister is empowered to pay the finder a reward not exceeding the prescribed amount of $50,000. The following examples give some idea of the range. The finders of HMS *Pandora* (1791) were rewarded with $10,000; the finders of the *Rapid* (1811) were given $30,000; the finders of the 'blackbirder' *Foam* (1893), sunk off the Queensland coast received $2,000; the finders of an unidentified whaler of around 1810 on the Rowley Shoals received $2,000; and the finders of the timber-trade barque *Contest* (1874) in Cockburn Sound were given $250.

Chapter 5 REFERENCES

1. McGimsey, 1972, p. 4.
2. Ryan, 1977, p. 25.
3. Roper, 1978, p. 6.
4. Major, 1859.
5. *Daily News*, 1963, p. 1.
6. Roper, pp. 49-135.
7. Lewis, 1964, *Hansard*, p. 2328.
8. Crawford, 1977, p. 32.
9. O'Keefe, 1978, p. 3.
10. O'Connell, 1965.
11. Crawford, 1977, p. 31.
12. Bolton, 1977, pp. 28-30.
13. Ryan, 1977, p. 24.
14. The New South Wales Heritage Act gives some protection in that State also.
15. Amess, 1983, p. 52.

6 Early East Indiaman Traders: Pre-Settlement Maritime Archaeology at the Western Australian Museum

The principal museums in Western Australia are all closely associated. Part of the intention of the Museum Act of 1969 was to rationalise the growth of museums throughout the State. Prior to that time the Tourist Development Authority and the Royal Western Australian Historical Society had sponsored some local and specialised museums and the Museum Board was increasingly called upon for assistance and advice in the form of technical services, loan of specimens and displays.

The Government wanted to prevent the proliferation of inadequate museums based on little other than a sudden surge of sentiment. The Act was designed to discourage haphazard development but at the same time encourage local authorities to open functional local museums concerned with preserving a record for posterity of the characteristics of the history and natural history of their region, and to display these collections for the interest and education of the community, particularly newcomers and visitors. This has ensured continuity of operation and the safety of collections, and enabled some museums to obtain the assistance of the professional and technical services of the Western Australian Museum.

So today there are local and branch museums at regional centres including Fremantle, Geraldton, Albany, Esperance and York. Displays of material from shipwrecks are to be seen at Fremantle, Geraldton and Albany, although the maritime archaeology and materials conservation staffs are located at Fremantle.

The Western Australian Museum is funded by the State Government but is not a government department. The Director is responsible to a Board of Trustees appointed by the Governor. The staff is broken into three basic divisions: Natural Sciences, including departments of crustacea, ichthyology, molluscs, entomology etc; Human Studies, including maritime archaeology, history, anthropology and archaeology; and Professional Services, including material conservation and restoration, display, education, library, publication, and local museums.

The Maritime Archaeology Department is responsible under the Commonwealth and State Acts for the management and excavation of shipwrecks in waters off Western Australia, and for the research and maintenance of the resulting collection. Fieldwork and the associated research work is divided into two areas: the pre-settlement Dutch wrecks and the seventeenth century *Trial* (run by the Department

Head, Jeremy Green), and the post-settlement period sites (run by a Curator, the author). As a large number of wrecks occurred in the post-settlement period, a continuous programme of inspection of sites is maintained. Within the office, laboratory and workshop a professional and technical staff of 13 carry out work on the collection and maintenance of field equipment.

The Maritime Archaeology Department also has an educational role. In 1980 and 1981 professional staff from the Department joined with specialists from the Western Australian Institute of Technology and the University of Western Australia to provide a one-year graduate diploma course in maritime archaeology based at the Institute. The course was intended to fulfil a perceived need for qualified maritime archaeologists in most Australian States, as institutions in those States became interested in pursuing the aims of the Commonwealth legislation of 1976. Seven of the course graduates (each course produces about 10 graduates) currently hold positions relating to maritime archaeology in Victoria, Tasmania, South Australia, Western Australia and Queensland. The course, which has a strong technical orientation, is not held every year, but only when a need is seen. It is being run again in 1986, for the third time.

The Department works in very close co-operation with the Materials Conservation and Restoration Department, because almost all material from shipwreck sites requires some treatment. Curator Dr Ian MacLeod currently runs both the conservation and restoration sections.

The Maritime Archaeology Advisory Committee makes recommendations to the Trustees on matters such as rewards and the status of particular sites. The Committee's membership is currently drawn from the University of Western Australia (an historian), Murdoch University (an historian), the Institute of Technology (a physicist), the Library Board of Western Australia (an archivist), the diving equipment retailers, the diving fraternity and the Museum. The membership enables a broad spectrum of community and expert opinions to be voiced.

The stimulus for a maritime archaeological fieldwork programme came from the knowledge that four early Dutch trading ships (the *Batavia, Vergulde Draeck, Zuytdorp* and *Zeewijk*), and one English trading ship (the *Trial*) lay off the coast. The wrecks of other early outward-bound East Indiamen are now known to lie off Europe, South Africa and elsewhere, but Western Australia pioneered the archaeological approach to this category of shipwreck sites.

1. The Batavia Wreck Excavation

The wreck of the outward-bound Dutch trading ship *Batavia*, lying in 5 metres of water, was discovered in June 1963. The wreck site lay on the south-west corner of the Morning Reef in the Wallabi Group of the Abrolhos. The main body of wreckage extended about 50 metres by 15 metres and was located in a slight depression, roughly the shape of a ship's hull, which may be coincidental, but is likely to have been caused by the ship settling into the reef. A thin hard crust of coral, held together firmly by living organisms, overlies a thick layer of loosely packed coral fragments. The grinding weight of the 42-metres-long wooden vessel could well have

The remains of one of the castaways from the *Batavia*. It has been suggested that the mark on the top of the skull is a sword cut, and that the skeleton is that of Andries de Vries. (Photo: Pat Baker)

punctured the crust before disintegrating. Water depth on the wreck varies from 4 to 7 metres and in bad weather booming Indian Ocean swells of up to 10 metres in height smash themselves onto the site.

The islets lying closest to the *Batavia* wreck provide scant human shelter. Small chunks of dead coral have been washed together by storms to a height just above sea level. The island group is 80 kilometres from the mainland, and remains uninhabited except during the annual rock-lobster fishing season. The multitude of sea birds are given some protection from industry and tourism by the provisions of wildlife protection legislation.

When we archaeologists first ventured across to the windswept islets we came to realise how little shelter they give from the elements; our tents were blown down and dinghies slipped their moorings. To work successfully on the wreck site required substantial island-based facilities. Fortunately, the potential of the *Batavia* wreck was appreciated in all quarters from the very commencement of excavation work. The State Government responded with generous funding which enabled the construction of a permanent base camp on Beacon Island, located just 2 kilometres from the wreck itself. Beacon Island, referred to by the *Batavia* wreck survivors as Batavia's Graveyard, was the scene of Jeronimus Cornelisz's worst excesses. Skeletons have been recovered from beneath the floors of fishermen's huts, and more are known to remain there. But there are no obvious signs of the temporary Dutch presence. Two stone constructions both date to the nineteenth century: one a surveyor's beacon, the other a low wall believed to have been built to shelter survivors from a later nineteenth century shipwreck called the *Hadda*.

The Museum's house, built in 1971, provided accommodation for six people. For self-sufficiency it was equipped with lighting plant, freezer unit, photographic darkroom, drawing office, radio communications, artefact storage and field conservation facilities. Water from the roof was collected in storage tanks for washing but the high content of sea-bird droppings rendered it undrinkable.

A jetty was extended out from the island into deep water so that large boats could come alongside for refuelling and loading. The Museum commissioned the 12-metre steel workboat *Henrietta* with twin diesel engines for safety. The stern of this vessel was fitted with a 2.5-tonne winch and a hydraulic 'A' frame, which gives it the capacity for lifting heavy objects such as cannon on board. The deck is large enough to carry semi-permanently the air-lift compressor, used for clearing sand from the wreck site, and the small hookah unit for supplying air to divers.

The Museum carried out four seasons of excavation between 1973 and 1976. My first dive on the *Batavia* left me with a lasting impression of the vivid beauty of the marine life; the unexposed wreck was less conspicuous. The site appeared roughly similar to the *Trial*, there being a haphazard scattering of 25 cannon, 9 anchors, concentrations of ballast bricks and building blocks, and a small scattering of lead fragments and ceramic shards. But two privately organised expeditions and the four conducted by the Museum have shown the archaeological value of the *Batavia*. Hidden beneath the coral was a section of the hull, comprising the after section of the port side of the ship and part of the transom. These are the earliest known substan-

The 11-metre Museum workboat *Henrietta* moored over the *Batavia* wreck site in marginal conditions. (Photo: Pat Baker)

An unexpected large wave overwhelms the *Henrietta*, leaving only the winch 'A' frame on the stern exposed. (Photo: Pat Baker)

The vessel emerges from the wave. The anchor has held, but staff are injured by flying glass. (Photo: Pat Baker)

tial remains of an East Indiaman. Built in the same year as the well-known Swedish warship *Wasa*, now raised and preserved in Stockholm Harbour, the *Batavia* is filling some of the gaps in our knowledge of early seventeenth century shipbuilding.

The significance of the *Batavia* wreck extends beyond its timbers. A striking feature of the VOC's merchant fleet was that it fulfilled a double function: the ships were expected to be able to fight as well as carry commercial goods. The military hazards of the East India voyages which sprang initially from the hostility of the Portuguese, later from the Dutch desire for monopoly, and throughout from the activities of pirates, made it absolutely essential for the Company to equip and despatch vessels well-armed and in convoy. A wide variety of military equipment was found on the *Batavia*. There were ordnance and firearms in the form of iron cannon, bronze cannon, two curious and unique cannon made up of sandwich layers of copper and lead over wrought-iron rings, several blunderbusses, and musket stocks. The number and variety of projectiles was astonishing—iron ball shot, bar shot, expanding bar shot, spike shot, canister and langrel (bagged scrap iron) shot, grenades and bomb shot, and a range of lead shot. Then there were cannon touch-hole prickers, copper gunpowder canisters, gunpowder measures of differing sizes, gunpowder scoops, a cannon muzzle plug, sword pommels, brass ferrules (end strengtheners) from pikes, a gunner's tallystick, and even a suit of armour, including the delicately gilded breastplates and leather fittings.[1]

The wreck has also proved to be a veritable warehouse of navigational equipment, illustrating materially the strengths, and some of the weaknesses, of early seventeenth century Dutch navigating. A semi-circular astrolabe and several circular astrolabes for taking altitudes and obtaining latitudes, the meridian ring from a celestial globe, a section of a Barents' arm from an Astrolabium Catholicum, a brass semi-circular protractor, brass dividers for pricking off distances sailed, and sounding leads, form just a part of the magnificent collection of navigational equipment found on the *Batavia*.

The first phase of the excavation was the lifting of 128 delicately hewn sandstone blocks, which weighed in all some 30 tonnes. Rope strops were fastened around each piece before it was hoisted aboard the *Henrietta*. Then the blocks, ranging in weight from 10 to 100 kilograms, were taken on shore at Beacon Island for cleaning, inspection and cataloguing, before transport across to the mainland at the close of the season. Many of them were found to have been numbered; inscribed by European stonemasons to indicate their position in the planned structure on assembly. The various components, which form two columns and a pediment when assembled, appear to have been originally intended for the water port of the Castle at Batavia. A contemporary engraving shows the water port in the early stages of construction, with scaffolding in place but the stonework yet to come. A slightly later illustration shows the gate complete, so it would seem that another set of blocks was successfully ordered and delivered to replace the one lost with the *Batavia* wreck.[2]

As we systematically removed the sandstone portico a broad expanse of timber flooring emerged. The planks had been pinned to the seabed by the combined weight of the portico, a number of cannon, and a large quantity of solidly concreted cast-

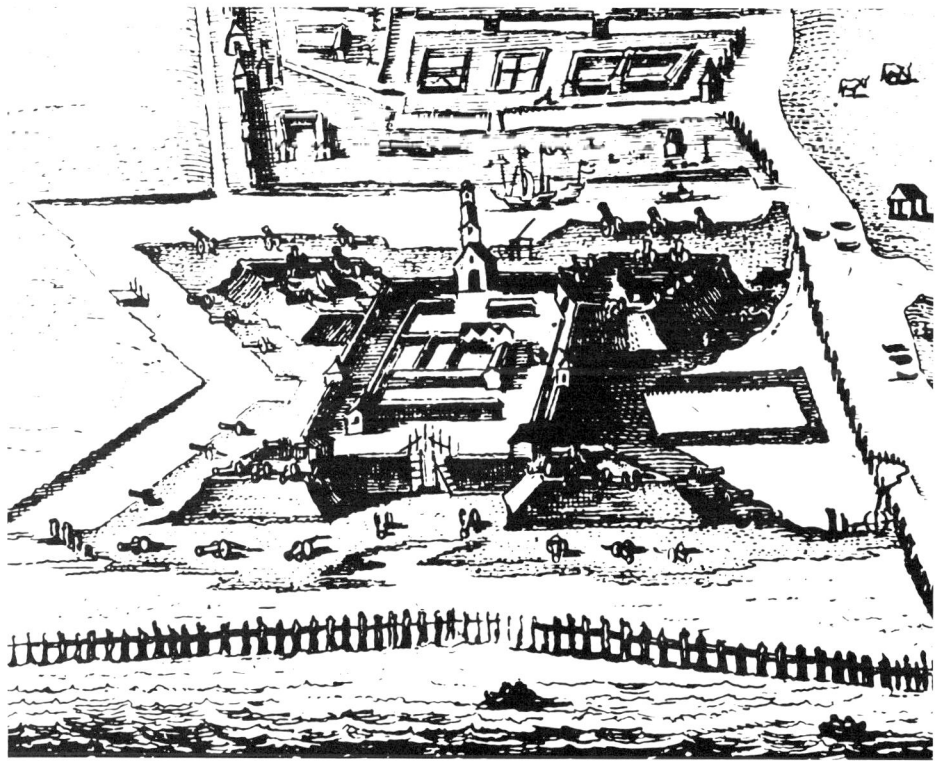

The Dutch fort at Batavia (present-day Djakarta) in 1629, showing scaffolding around the water port in the foreground. (Photo: Western Australian Museum)

iron shot comprising the contents of the stern shot locker. The cannon—iron, composite and bronze—were quickly floated using lifting balloons which were filled with air expanding to look like giant bloated mushrooms, waving with the tide as they strained to escape the seabed, and then popped gently up to the surface. But the thousands of cannonballs were a more difficult problem. They were cemented together so firmly that each ball represented hours of laborious chipping with a geo-pick and hammer. Even then the end product was likely to disintegrate into a carbon cloud, because over the years much of the metal strength of the individual balls had leached into the surrounding concretion. This concretion, which struck sparks from steel on dry land, had become hard as iron, while the graphitised cannonballs had gone soft and brittle. For shot of 0.14-metre diameter, with a theoretical weight of about 11 kilograms, actual weights after leaching ranged down to 3 kilograms, without any alteration of shape.

After giving due consideration to all the alternatives, Jeremy Green decided to use small quantities of explosives to try to fracture the concretion.[3] The decision was not taken lightly, for the Museum was mindful of the appalling wanton destruction done with explosives before protective legislation was introduced. In this case, however,

The sandstone blocks, thought to have been intended for the Dutch fort, were raised from the *Batavia* wreck and reconstructed in the Maritime Museum. (Photo: Pat Baker)

the technique was used with the delicate precision appropriate to archaeological excavation. A series of experiments were carried out to see if acceptable results could be obtained. Charges of 2-gram Metabel explosives were used. The explosive, linked to a cordtex train and electric detonator, was placed on a heavy rubber mat over the area to be fractured. To be acceptable the technique had to produce results which caused less damage to artefacts and at the same time increased the pace of the work. In both respects the method proved successful. The shock waves caused a shearing effect at the interface between materials of different hardness, and by this means the balls were freed. After the charge had gone off, the fragments of the shattered concretions were quickly chipped and brushed away, and the cannon balls, undamaged, fell loose. With the cannon balls removed the recovery of the ship's timbers could be completed.

Excavation revealed a section of the port side of the ship extending 12 metres by 6 metres. This included parts of the sternpost, stern, and the side of the ship. The upper works, the starboard side of the ship, the bows, and the keel, had all disappeared entirely. Surviving load-line marks on the sternpost are graduated in Amsterdam feet and the numerals range from XVo to XX: the 16 to 20 feet marks above the base of the keel. These make it possible to estimate where the keel should have been had it survived. The upper limit of the surviving section was terminated at the position where the roof of the first gundeck once was. The *Batavia* had crashed bows first onto the shelving reef, and it was to be expected that the stern section of the ship, lying in deeper water, would be the best preserved.

Among the intact timbers was a gunport, pierced through a wale. On the outside of the port the corroded remains of the iron hinges could be seen, with the ring by which the door was opened upwards and outwards. It had been hinged in this manner to avoid being dashed inwards by the weight of the water upon the ship's side when she heeled under heavy gusts.

The wooden walls of the *Batavia* were extremely thick, and built up in a complex manner. The frames, or ribs of the ship, were covered on the inside with ceiling planking. This in turn was covered with a thin timber skin to shield the ceiling from damage by heavy cargo. Outside the frames was another heavy sandwich of four layers of planking, consisting of a thin inner skin, the inner strake layer, the outer layer of strakes and a thin layer of soft-wood sheathing. The sheathing was attached to the strakes with a matrix of large headed nails, driven through a layer of tar and cow hair, and closely spaced to produce an iron rust layer as a deterrent against ship worm.[4]

The *Batavia* was a 'spiegel' or flat-backed ship, typical of large vessels of the first half of the 17th century. The timbers raised from the wreck illustrate the defining characteristics of this type of vessel. Forming the shape of the side of the ship at the junction with the stern is the fashion piece, a massive timber supported by heavy knees and diagonal planking. Part of a stern gun-port has also survived.

Removing this extremely heavy and complex structure from the seabed in ocean swell conditions was quite a challenge. The ever-present surge on the site precluded raising it intact, so as each layer of timber was exposed it was labelled with code

The port side (left of ship looking forward) of the transom of the *Batavia* is exposed during excavation. (Photo: Pat Baker)

numbers and systematically photographed with a two-dimensional scale in each shot. Thus a single composite photograph or photo-mosaic was built up. This layer was then removed plank by plank and the same procedure repeated on the underlying layer, resulting in a series of photomosaics of overlapping layers.

Frames and knees were removed from the structure by gently easing them off the attached planking. Many of the planks were too long to be recovered in one piece, some being over 10 metres long. The winch had the capacity to lift them but at that length the planks would break under their own weight once lifted from the water. The oak was too hard for sawing by hand so an air-powered chain saw was used to produce manageable 3-metre lengths. The sawdust discoloured the water but the resulting edges were clean cut.

On board the *Henrietta* the wood was continually hosed during the short voyage to Beacon Island, where plastic-lined storage pits filled with seawater and fungicide had been prepared in anticipation. More measurements were taken on the island. Each surface of every timber was re-photographed and traced on to plastic sheeting. Then at the end of the excavation season the pieces, still wet, were wrapped in polythene sheeting and taken to the Department of Materials Conservation in Fremantle. Some 30 tonnes of timber arrived at the laboratory in this way.

The most outstanding example of the preservation of an historic wooden shipwreck is that of the Swedish warship *Wasa* at Stockholm. After being raised to the surface virtually intact she was towed on her own keel to the dock, slowly entering one evening, with a slight list to port. The marvellous state of preservation of the *Wasa* can be attributed partly to the lack of wood-boring teredo worm in the brackish Baltic waters, and partly to the circumstance that the vessel was commencing her maiden voyage when she heeled over and sank, undamaged, to the muddy bottom of a deep harbour, rather than running aground in shallow turbulent waters.

The *Wasa* approaches the ideal ship burial situation: a vessel built of new wood, creosoted or treated against biodeterioration, which sank rapidly into deep sediment with a low oxygen level environment, below the level of anaerobic bacteria. As the Canadian conservator M. -L. Florian has put it:

> The wood treatment would kill any fungi or bacteria which might have been carried down on the surface of the wood and the depth of the sediment would prevent any infiltration of organisms from the surface. The marine borer cannot attack wood in the sediment and would not be present in the wood because it takes at least 10 days submersion to precondition the wood prior to marine borer attachment. In this-environment, biodeterioration may not occur but waterlogging is inevitable. Waterlogged wood because of the physical changes caused by the water at the molecular level and the chemical changes by the water leaching of soluble components of the tissue will still require conservation treatment to prevent distortion on drying.[5]

So even on the *Wasa* there are enormous conservation problems. The hull was completely waterlogged, and if left to dry without treatment would have shrunk and cracked irreparably. The outer surface of the timbers had been affected by soft rot and had lost mechanical strength. Treatment of the *Wasa* hull involves the replacement of the salt water from the cellular structure of the wood with water soluble polyethyleneglycol or P.E.G., a plastic-like substance whose consolidating function could be likened to the sap in a tree. The preservation time for the 1,250 tonnes of timber comprising the *Wasa*, where the P.E.G. is applied by spraying, is estimated to be 23 years. In Germany the hull of the Bremer Kogge, which sank in the river Weser about 1380, will be treated with P.E.G. for 20 to 30 years in a large vat, while in England the *Mary Rose* hull is currently being sprayed with cold water and fungicide.[6]

The loose timbers of the *Batavia* have been preserved in vats, using P.E.G. of varying concentrations for penetration. Some problems of shrinkage on drying after treatment were experienced, but the conservators at the Museum have benefited from the knowledge of the *Wasa* experiments, and the treatment time has been drastically shortened. The last batches of *Batavia* timbers came out of the vats in 1986, but they require a permanently humidified environment.

The surprising thing about the *Batavia* timbers is that any survived at all on the seabed. Most of the ship was torn apart by the physical force of heavy seas soon after she struck the reef. The survivors noted that the keel itself had broken from the site and drifted to a nearby islet. Besides the turbulence the organic conditions were against survival of the timbers. Although a new ship, on her maiden voyage, she had spent 7 months at sea before striking the reef: ample time for worm to establish itself. The warm, shallow, oxygen-enriched coral reef waters of the Abrolhos provide an ideal environment for marine borers and a host of other destructive organisms. Few of the factors generally cited by conservators as being favourable for preservation existed on the *Batavia* wreck site.

Perhaps the very turbulence which rapidly destroyed the upper works and the entire starboard side of the hull, and even wrenched free the keel, was also responsible for the preservation of the remainder. The timbers lying pinned beneath the sandstone portico and the heavy cannon were rapidly and completely buried as each successive storm brought more coral sand and coral fragments to filter into the crannies between the blocks and form an 'airtight' protective mound over the structure.

As the timbers are conserved they are being re-assembled in a humidified section of the old Fremantle Commissariat. This mid-nineteenth century warehouse was built of local limestone during a period when convicts were being transported from England, providing a ready supply of labour. Restored as the Western Australian Maritime Museum, the building retains its atmosphere of austere roominess, and provides ample space to house the 6-metre height of the *Batavia* remains, as well as the 7.5-metres-high sandstone portico (surmounted by a watchful lion), now fully erected in the same room, and a variety of other displays about the vessels wrecked on this coast. Museum visitors are provided with observation galleries at ground and first-floor levels, so that they can watch the archaeologists re-building the salvaged section of the *Batavia*, a dynamic aspect of the creation of a museum exhibit.

Thirty tonnes of timber was recovered from the seabed: about one-seventh of the ship's hull. This included part of the transom and the port side at the stern. The support system for the reconstructed timbers will not be visible from outside the hull. A steel framework will support the weight of each individual plank as if it were set on a shelf.[7] The inside of the hull will be available for display of artefacts relating to the ballast, cargo and shipboard life. The work is scheduled for completion in 1987.

2. The Vergulde Draeck Shipwreck Excavation

This wreck was discovered by spearfishermen in 1963. It lies in 7 metres of water on a limestone reef 5.6 kilometres off the coast about 100 kilometres north of Perth. The reef is honeycombed with limestone solution tunnels, originally coastal landforms but now submerged and heavily eroded. When first found the most obvious features of the wreck site were a mound of small ochre-coloured Dutch ballast bricks, a heap of elephants' tusks protruding like errant scythes from the seabed, 14 iron cannon and 6 anchors partly obscured by kelp weed.

The excavation of most of the remains of the *Vergulde Draeck* was completed in a single season, lasting from January to April 1972. In many ways it was a testing ground for maritime archaeology in Western Australia. It was the first substantial excavation attempted by the project leader, and the first full-scale expedition organised by the Western Australian Museum. Circumstances dictated that the work be undertaken without delay: the site had legal protection but it was common knowledge that private skindivers (some using explosives) were looting the site for the silver coins and other items on board the vessel when it sank. The results of the excavation season on the *Vergulde Draeck* gave impetus to an active excavation programme which has continued on to other West Australian sites without abatement.

A hut was erected at the small rock-lobster fishing village of Ledge Point, 12 kilometres from the wreck site. The principal workboat was a 6.5-metre plywood vessel: quite unseaworthy but the only vessel available at the time.

One of the most difficult problems encountered in excavating the *Vergulde Draeck* wreck site was finding a suitable mooring position for the workboat. On the seabed, diving conditions were better than on the *Batavia* wreck, because divers on the deeper *Vergulde Draeck* site worked under the protection of a small submarine cliff which directed the force of the swells upwards, whereas on the *Batavia* site the diver was washed forwards and backwards the full wave length of the unobstructed swells moving up a gently sloping seabed. But on the surface the *Vergulde Draeck* site was the more formidable. A heavy mooring was established for the workboat 100 metres seaward of the wreck, and diving was abandoned whenever a wave broke more than 75 metres seaward from the site.

An airlift was used to suck away sand covering the artefacts. The spoil was directed by the airlift to the surface where the waves carried it forward, over the reef and clear of the site.

As the sand was removed vast numbers of ballast bricks emerged. Eight tonnes of these were bagged in 20-kilogram lots and floated from the site with lifting balloons on the calmest days. Other more delicate artefacts were found both above and below the bricks.

The surface layer of sand showed clear indications of having been previously disturbed by pot-holing souvenir hunters, but even the deepest layers had suffered badly from turbulence after the ship had first struck the reef. The *Vergulde Draeck* broke up almost immediately upon striking. No ship structure was found, so it would seem that the bottom of the ship broke open, the heavier cargo was swirled into depressions in the reef, and the ship's timbers were thrown over the reef to float away.

Despite the disappointing absence of many meaningful pieces of the ship itself, the wreck is an interesting agglomeration. A wide range of accurately dated domestic items, provisions and cargo were found on the site. The individual artefacts in this collection serve to help in identifying and dating other sites.

One of the most interesting categories of material are the bulbous Rhenish stone-

Carl Benko uses a vibrator tool to clean coral from a large mound of ballast bricks from the *Vergulde Draeck* (1656) wreck. (Photo: Vera MacKaay)

ware jugs known as beardman or bellarmine jugs because of the leering face which appears on the base of the necks. About 50 of these jugs, most of them bearing medallions and exhibiting the characteristic salt-glaze 'orange-peel' effect, have been recovered undamaged from the wreck by the Museum. They constitute one of the world's largest collections of dated beardman jugs and are excellent material for analysis, particularly when viewed together with the thousands of additional beardman shards from the same site. The masks range from semi-naturalistic to animalistic and the quality is equally variable: two of the jugs have holes which obviously represent manufacturing flaws.[8]

The array of clay smoking pipes is another interesting area of study. Eight types were found and the collection includes a box of 250 pipes all of one type. The complete pipes from the box were 38 centimetres long. It is a rare situation to have so many pipes in good order and made by the same person at the same time. One of the dating methods currently in vogue is based on statistical measurements of bore diameters of pipe stems, which are thought to have changed gradually through the seventeenth century.[9] Yet this large collection off the *Vergulde Draeck*, all from the same date and maker, proved to have two stem hole diameters.[10] Another pipe, of

Clay smoker's pipes from the *Vergulde Draeck* site are processed by archaeologist Jeremy Green. (Photo: Jeremy Green)

brown body ceramic with complex decoration, appears to be Southeast Asian in origin, and has prompted the suggestion that pipe-smoking in Southeast Asia pre-dated that in Europe.

The coinage throws light on seventeenth century numismatic practices. About 8,000 coins, mostly Spanish silver reals minted in Mexico, were raised during the excavation. One of the questions examined was that of how long it took for coins minted in Mexico to reach Batavia, via the various government and trading channels. The *Vergulde Draeck* sailed from Texel, in the Netherlands, for the East Indies in October 1655. The most recently dated coin found on the wreck was a Potosi (Bolivia) 4-real piece of 1654. That piece should have been carried by llama and galleon to Panama to await the January 1655 treasure fleet to Spain. But the fleet for that year was delayed for over 12 months, so despite all the regulation of the traffic, smuggling may have been the only possible means of its transport to Holland in time for the *Vergulde Draeck's* October 1655 departure.[11]

A related issue is the hoard of about 40 silver coins, found with the remains of a small chest in sand hills on the shore adjacent to the *Vergulde Draeck* wreck site in 1931, and consisting of Japanese coins called Mameita-Gins, and a number of ducatons and half ducatons ranging from 1637 to 1655. The 1655 ducaton was minted at Brussels in the Spanish Netherlands, a short distance from Texel. So the question which arises here is whether the issue date of the 1655 coin was before or after the *Vergulde Draeck*'s departure date. If it was after, then clearly the hoard was not left on the beach by a shipwrecked Dutchman from that vessel, but someone else who came later.

Other material with the potential for study includes the elephant tusks (from African elephants); a tool box, raised complete and then excavated in the laboratory to extract the variety of tools; and the bones, consisting of shanks, ribs and shoulders of what was once salt beef and pork, and which might tell us more about the domestic animals of that period. There were even bones (identified by their skulls) of *Rattus rattus*, the common household rat. Unlike the *Vergulde Draeck*'s master, Pieter Albertsz, these rats had gone down with the ship. But it is certain that others would have drifted ashore clinging to spars, becoming the first of a wide variety of vermin introduced to this continent from Europe. If rabbits, goats, cats, foxes and starlings had reached the shore at that time to wreak their havoc upon the environment of Australia perhaps the botanist Sir Joseph Banks, a century later, would have been more hesitant in his report about the potential of the country.

3. Other Early East Indiaman Shipwrecks

One of the characteristics of archaeological sites is that they retain some elements of mystery even after excavation. Indeed, a lot of the mysteries only begin after excavation. Normally, when a large-scale excavation is attempted, particularly in mobile sand environments, it is impossible to be certain that some material has not been missed, and it is rarely the intention of the archaeologist to remove everything anyway. In the case of the *Batavia* and the *Vergulde Draeck*, anchors and cannon are known to remain on the sites and doubtless there are pockets of unknown material that were not located during excavation, perhaps 200 metres inshore for example. But most of the material has been raised, to become part of a Museum collection.

Archaeological interest has been directed towards the remaining three known early East Indiaman wrecks, but excavation work has been limited. The *Zeewijk* wreck on the Pelsaert Group of the Abrolhos has received most attention.

Relics from the *Zeewijk* survivors' island camp-site were found in 1840 by the surveyor John Stokes of the *Beagle* (Charles Darwin had left the *Beagle* by that time). Stokes took a small bronze gun back to England with him. Then in the 1890s, when guano (a natural fertiliser consisting of sea-bird excrement) was being mined on Gun Island, a collection of bottles, cooking utensils and other objects was assembled. Most of this material is now housed in the Museum. In the mid-1950s the Royal Australian Navy was led to the wreck site by a rock-lobster fisherman. Navy divers raised a number of cannon from the lee side of the reef, but it was not until 1968 that skindivers first swam over the reef-top to find more cannon and anchors on the seaward side where the *Zeewijk* had first struck and slowly disintegrated. This resulted in an ex-gratia payment of $1,000 by the Museum to the finder.

Maritime archaeologists from the Museum paid a brief visit to Gun Island in 1974

Guano diggers unearthed this collection of *Zeewijk* (1722) material at the turn of the century, and a photographer recorded the assemblage. (Photo: Western Australian Museum)

to assess the archaeological potential of the island occupation site and the wreck site, and carried out extensive surveys of the material in 1976, 1977, 1978 and 1979.[12]

The *Zeewijk* is perhaps the least significant of the East Indiamen found wrecked off the West Australian coast. Being the most recent of these wrecks, it belongs to a period about which we already have considerable knowledge. More records about shipbuilding, trade and ways of life have survived from the early eighteenth century than from the early seventeenth century when the *Batavia* went down. A site contemporary to the *Zeewijk*, located on British shores, is in a much better state of preservation.

The Dutch East Indiaman *Amsterdam* set out on her maiden voyage in late 1748 (22 years after the *Zeewijk*), fully laden with a rich cargo and more than 300 people aboard. She encountered a gale in the English Channel and was beached on tidal flats near Hastings in Sussex. The vessel was buried in the soft sand so fast that there was barely time to unload her treasure. So now, three-quarters complete, the ship is an almost untouched storehouse lying in a position easily accessible to archaeologists. The site was located in 1969 but so far the funds necessary for full excavation have not been made available by either the British Government or interested Dutch groups.[13]

The more positive approach by the Australian State and Commonwealth Governments to the *Zeewijk* wreck has enabled archaeologists to establish (despite adverse environmental conditions) what is and is not there. No substantial structure has survived from the *Zeewijk*. The gently shelving reef offers no convenient depressions or pot-holes for protection of sections of the ship from the surf. Indeed, the site of the original striking has never been accessible to sustained survey or excavation because of the continual turbulence and heavy cross currents which make even a straightforward tape and compass survey a formidable task.

The Museum's attention has been directed towards the material thrown over the reef into the sheltered shallows, and the relics on the island some 4 kilometres away. The cargo items found on the leeward side of the reef were fragmented and widely dispersed. In collecting the material, which consisted principally of glass bottle and beardman jug fragments, distribution charts were drawn showing how successive storms had thrown material in a fan pattern extending several kilometres. The large collection of shards is suitable for statistical analysis. Gun Island, where the survivors camped, initially appeared to offer some archaeological promise, but the majority of this landscape was scoured to bedrock and beyond by the guano diggers of the 1890s, leaving only a narrow strip of sandhills on the west side of the island where artefacts might still be undisturbed by man. And here, any stratigraphy has been destroyed by the thousands of burrowing seabirds, a type of petrel popularly known as muttonbirds, which nest annually. Magnetometer surveys and numerous test-holes have yielded a far less impressive collection than the picks and sieves of the early guano miners.

Wreckage from the *Zuytdorp* was found in 1927 lying ashore at the foot of cliffs about 60 kilometres north of the Murchison River. A stockman on Tamala pastoral station found glass bottles at the top of the cliffs and lower down among the rocks the carved wooden figure of a woman, which had decorated the supports below the stern windows of the *Zuytdorp*. In 1954 a private expedition found more relics along the shore-line, including a number of coins which confirmed the identification of the site and showed that some of the crew had camped for a while on the cliff-top. Thinly buried ashes at the peak possibly indicate a signal fire, desperately tended by the survivors in a fruitless attempt to attract the attention of a passing Dutch ship.

The wreck itself lies in shallow water, 3 to 6 metres deep, against a shelf of rock at the base of the cliffs. Lying immediately below the shore breakers, the site presents extremely difficult diving conditions, and the wreck was not examined by skindivers until 1964. In 1971 Museum staff raised 3,500 coins, a bronze cannon, a ship's bell and other material.

The archaeological potential of the *Zuytdorp* is linked with the turbulent underwater conditions at the site. The ship was presumably smashed to pieces almost immediately upon striking the cliffs: otherwise the silver coins (the most valuable cargo item) would have been carried ashore by the survivors as their first priority. This suggests that little else would have been taken from the wreck by the survivors, a positive factor in terms of archaeological potential. But, on the other hand, those savage seas which smashed the ship when she first struck, have dispersed and

An observer stands several metres from the turbulent *Zuytdorp* wreck. (Photo: Susan Green)

destroyed the timbers and contents of the ship to such an extent that little which is not durable can be expected to have survived. Even a bronze cannon raised by the Museum had been halved longitudinally by wave action, exposing a cross-section view of the interior bore. Sand particles, dashed backwards and forwards by the tenacious fury of the waves had, over the centuries, worn through the big gun with an action like that of the prisoner's file on his cell bars. It may be that the lower layers of coins, cemented to the seabed and protected by the seas both from the survivors of the wreck and latter-day would-be looters, will prove to be the most interesting archaeological material to be found on the *Zuytdorp* wreck. The consignment comprised most of a special 1711 minting of 100,000 guilders value in schellings and double stuiver coins struck at the Middelburg Mint.

The Museum has a limited archaeological salvage programme in operation on the *Zuytdorp* wreck. The first objective is to excavate the areas where the silver coins are concentrated, because it is likely that these areas will otherwise be destroyed by looters. Unfortunately, of all the material on the site, the coins lie closest to the cliff directly under the heaviest surf. Diving is only possible in ideal weather conditions, which occur about 10 days each year. A field camp was for a time established on the remote cliff top and a watchkeeper radioed site weather conditions to the Fremantle office. When conditions were judged to be satisfactory for diving, a team flew to the site by a small aircraft which landed on a rough cliff-top air strip graded from the scrub for this purpose by the Nizam of Hyderabad, owner of the Murchison House pastoral station. A small flying fox was erected on the cliff-top, leading down to the area where excavation was intended. Divers, wearing safety helmets, were lowered into the water on the flying fox. Now the wind has scoured sand from the air strip, making it dangerous, and the salt spray has made the flying fox unusable.

In the 1960s, before excavation work began on the shallow, swell-affected wrecks off Western Australia, it was a widely held view among archaeologists (most of whom had had their experiences on deep-water wrecks in the Mediterranean) that shallow sites would not yield any significant remains because of the destruction wrought by the turbulence. The West Australian experience has shown both that substantial quantities of material can survive on such sites, and that they can be excavated according to archaeological principles. However, the *Zuytdorp* site goes close to the limit.

A wreck believed to be the *Trial* was discovered on the Trial Rocks in 1969 by a group of Perth skindivers who had gone to the area in search of the ship. Under the terms of the Museum Act an ex-gratia payment of $2,000 was made to the group, and in June 1971 a Western Australian Museum expedition sponsored by M. G. Kailis of Gulf Fisheries surveyed the site. Before commencing work on the site Museum staff had to remove unexploded charges left by inept looters. Then a series of overlapping photographs was taken by a diver swimming over the site. The photographs were assembled to form a photo-mosaic and from this a rough plan of the site was drawn.[14]

The wreck has suffered badly from the elements. Heavy swell conditions and a powerful underwater tidal surge have long since carried away all exposed timbers.

Indeed it is possible that on the very day the *Trial* struck, after the bow section fell apart, the bottom fell out of the ship, allowing the superstructure of the vessel to drift away to sink in deep water. This would have left on the reef only the hastily jettisoned cannon, and that material which lay immediately above the keel, including the ship's ballast and anchors which had been stowed below. The main part of the wreck site, lying in less than 6 metres of water, occupies an area about 30 metres square, containing cannon, anchors and large river-worn ballast stones. Very few small artefacts can be seen, other than lead shot and scraps of sheet lead.

The principal significance of the *Trial* wreck is related to its age. It is the earliest known ship to have been wrecked in Australian waters. The *Trial* belongs to a time when the towering poop decks and fore-castles of the 16th century were fast disappearing, and when the four-masted ship was being replaced by the three-master. Few shipwrights' drawings of the period exist, so paintings are the only common means of tracing major structural changes. The structure of the *Trial* is not indicated in any of the available paintings or documentary evidence, but the condition of the site itself will probably limit the potential for obtaining any new knowledge of shipbuilding from this source.

The *Trial* is the only known seventeenth century English East Indiaman to have been wrecked on the Australian coast. Was the cargo principally limited to metal goods and cloth, or was there some variety of finished goods on board? Were the English by that time mass producing glass bottles in sufficient quantities to export large amounts of beverages to Asia by this means, or were they still confined almost entirely to the use of stoneware containers, or bottles imported from the Dutch? Comparison between the artefacts from an English East Indiaman and a Dutch East Indiaman of the same period might lead to some interesting conclusions regarding the different economic policies of the two countries.

Many fundamental questions about the *Trial* wreck site await exploration. Even the identification, in the absence of firmly dated artefacts, remained very tentative until late 1985, when a cannon, recently excavated from the site, was de-concreted to reveal a weight mark in imperial measurement, and an embossed marking which could represent the trifoliate leaves of a shamrock, the national emblem of Ireland. The cannon therefore may be indicative of an early English ship. It appears likely to be the *Trial* because its location seems to correspond with the records of the survivors, and because it has a number of large cannon. But other early vessels could have been lost without trace in that isolated area. Excavation offers the only way of definitely establishing the question of identity, but further work on the site will be expensive because of the isolation. The Trial Rocks lie 110 kilometres from the nearest mainland coast, to the north west of the remote Monte Bello Islands where the British Government exploded an atomic bomb during the 1950s. When the Museum set up its 1971 expedition camp in an observation block-house left by the British, the islands were still a restricted area because of lingering contamination.

Chapter 6 REFERENCES

1. McGrail, 1974, pp. 157-160.
2. Green, 1977, p. 68.
3. Green, 1975, p. 59.
4. Green, 1975, p. 49.
5. Florian, 1977, p. 135.
6. Barkman, 1977, p. 124.
7. Hundley, 1983, p. 225.
8. Green, 1977, p. 126.
9. Oswald, 1975.
10. Green, 1977, p. 162.
11. Simpson, 1980, p. 14. See also Wilson, 1977, pp. 293-339.
12. Ingleman-Sundberg, 1977, pp. 225-232. The work was funded by the Australian Research Grants Scheme.
13. Marsden, 1974.
14. Green, 1977.

7 Post Settlement Maritime Archaeology at the Western Australian Museum

From the time New South Wales became a convict settlement in 1788 vessels began to specifically visit the Australian continent for reasons of business, rather than to simply pass by to ports in Asia. This new and rapidly expanding traffic resulted in many more shipwrecks, particularly in areas where ships' navigators were presented with new and uncertain routes.

Western Australia was first occupied by Europeans in 1826, and the Swan River Colony began in 1829. Over 1,000 vessels are recorded as having been wrecked on this coast during the first century of settlement. The archaeological approach to these sites must be different to that adopted for the five earlier Dutch and English wrecks. Those earlier sites automatically assumed a certain degree of significance in the public mind because of their age and they were excavated because of a clearly perceived threat to each site. But many of the later sites—those of the Colonial period—are also of significance, and indeed have greater relevance to the Australian community. They too are under threat of looting.

A State-wide management and research programme, encompassing elements of what today many would term cultural resource management, has been developed for the many Colonial period shipwreck sites. This includes the compilation of background information into a register of all vessels recorded as having been lost, the inspection of each wreck upon being first located, arranging for the protection of significant sites under the legislation, marking of sites with plaques giving information about the vessel and its status, conducting pre-disturbance surveys on some sites, and, less frequently, selecting sites for excavation on the basis of archaeological potential and historical background. The programme commenced in 1964 with the compilation of a comprehensive sites' register, a very substantial undertaking in such a large State. The register was complete for the earlier sites by 1980.[1]

During this time of a rapidly increasing skindiving population with greater mobility and spare time, and easy access to all the necessary equipment for salvaging historic shipwrecks of the Colonial period it would be easy to argue for a speedy Museum programme of recovery, from as many sites as possible. Such a course of action, however, might well save many individual objects but at the expense of the greater knowledge and understanding of the past which can be achieved through a well-prepared approach to excavation.

Any excavation should be done in accord with the spirit of the International Charter for the Conservation and Restoration of Monuments and Sites (ICOMOS) which seeks to establish moral and technological standards for the monumental heritage. It should add substantially to a scientific body of knowledge, and be consistent with the conservation policy for the site.[2] An adequate research plan is necessary to ensure that these objects are achieved.

In normal circumstances an archaeologist will ask questions about aspects of the past and then seek appropriate sites to be used as information resources. If one of the sites under his management is suddenly placed under threat of destruction, however, he may be called upon to quickly save what he can of the artefacts, and in such circumstances his research findings may be very limited.

Extensive excavations (involving the removal of loose artefacts but not the intact ship's structure) have been or are being conducted on three Colonial period shipwreck sites in Western Australia, these being an unidentified whaler wrecked around 1810, the outward-bound American China-trader *Rapid* wrecked in 1811, and the ex-slaver *James Matthews* wrecked in 1841. More limited excavations have been conducted on the regional trading vessels *Cumberland* (1830), *Elizabeth* (1839), and *Lancier* (1839), the British emigrant barque *Eglinton* (1852), the regional trader *Lady Lyttleton* (1867), the coastal steamer *Xantho* (1872), the German-built barque *Hadda* (1877), and the colonial whaler *Star* (1880). Salvage archaeology projects have been carried out ahead of harbour works at Bunbury, Jervoise Bay, Careening Bay and Fremantle.[3]

Surveys have been carried out on many other sites by Museum staff, members of the Maritime Archaeology Association, and several diving clubs involved in locating sites, in particular the Living Water Diving Club.

The many sites which are not now being excavated will remain for the interest and enlightenment of future generations if divers develop an attitude of cultural resource conservation—something akin to the natural resource conservation attitudes widely displayed by divers now to the fish life and to marine mammals such as whales, but with the realisation that archaeological sites are a non-renewable resource.

1. The Excavation of an Unidentified Shipwreck

Identifying a shipwreck site is crucial if the site is to be placed precisely in its historical context. While a shipwreck of the historical period remains unidentified it is difficult to frame substantive questions for examination during excavation. Conversely, once a site has been identified the known background can frequently be employed to develop a variety of further questions about the ship and the men who lived on board. The shipwreck found on Mermaid Atoll is an example of an as yet unidentified site, and as such poses a problem for the archaeologists involved.

In 1979 a charter boat party saw wreckage on the exposed rim of Mermaid Atoll as they motored by in deep water. The Museum subsequently made arrangements to inspect the site. Wreck inspection is the standard procedure which establishes the position, condition and likely archaeological potential of each shipwreck site after it is reported as found.[4] Guided by the finders, Museum inspection staff located the

The small sand cay on Clerke Atoll, some 260 kilometres offshore, served as expedition headquarters for an underwater excavation in 1983. (Photo: Pat Baker)

wreckage (two Admiralty pattern anchors with wood stocks) on the western side of the Atoll, some 300 kilometres west of the port of Broome. A wreck was soon located in nearby shallow water on the outer rim. Cannon, a third anchor, and whalers' trypots (used for cooking whale blubber to produce oil) lay partly concealed by heavy coral growth in a narrow gully. The artefacts indicated that the vessel had been wrecked before the first settlement was established in Western Australia in 1829. It was apparent even then that the site represented an early European whaler. As such it offered the opportunity to learn more about early South Seas whaling, even without precise identification of the ship.

To find out more about the wreck, expeditions were sent to the site in July 1982 and August 1983. The Museum was supported with funding and volunteers by the American organisation Earthwatch. A base camp was established on a small sandy cay 25 kilometres south of the wreck site, and the 20-metre charter boat *Nellie Melba* moored over the wreck. The tide range of some 5 metres combined with ocean swells to make survey and excavation difficult. The highly oxygenated water had encouraged coral growth up to 1.5 metres thick over some artefacts.

It is now apparent that the wreck is a British or European whaler of several hundred tons, lost in the first two decades of the nineteenth century.[5] Artefacts on the site include three whaler's trypots, a blubber hook, bricks from tryworks, several copper coins, four anchors, seven 4-pounder cannon, and iron deck supports.

Clerke Atoll, some 25 kilometres from an unidentified whaler shipwreck, has a tide range of some 5 metres, but between tides the channels at left provide access to a sheltered boat anchorage.
(Photo: Graeme Henderson)

The artefacts are being analysed in conjunction with archival information in an attempt to identify the wreck. Some 1820s charts of the area indicate that a vessel named *Lively* was lost there.

One vessel of that name was owned by Daniel Bennett, who operated one of the largest British South Seas whaling fleets. The archival work has centred around testing the hypothesis that the wreck is Daniel Bennett's 240-ton whaler. This was a two-decked, ship rigged vessel, which carried ten 4-pounder guns. It was built of wood, iron bolted and sheathed with copper over boards, as was the wreck on Mermaid Atoll. Bennett's *Lively* was built in France in 1787 as the 14-gun naval cutter *Abeille*, but was captured as a prize by HMS *Dryad* of the British Navy in 1796 and subsequently renamed. Although the evidence appears to point to the wreck being that of Bennett's *Lively*, further work on the site will be necessary before a definite identification can be given.

The part played by the Americans in South Seas whaling is well known, but little has been published on the British contribution, and the available archival sources are sparse. The whaling equipment, the ship's armament, the construction of the vessel, navigational equipment, crew's possessions—all of these will, upon full excavation and positive identification, provide research material about the British South Seas whale fishery.

The history of the western third of the Australian continent has been well documented from the time of the first settlement in 1829. But for the period leading up to the first settlement—the 1780s through to the 1820s—very little is known about the activities of the various European nations in the waters off Western Australia. The exceptions are the official scientific expeditions sent out by Britain and France. Thus, every shipwreck of that period has the potential to fill some of the wide gaps in our knowledge of European activities and contacts. A second shipwreck falls into the period—the American China-trader *Rapid*.

2. The Rapid Wreck Excavation

In 1978 a spearfishing group found a shipwreck at Point Cloates, 1,100 kilometres north of Fremantle. The finders reported that a large mound of silver coins lay clearly exposed on the seabed, so for reasons of site security an expedition had to be mounted immediately. The overriding question asked at that time was—what ship is it? During three seasons of excavation the ship's timbers were surveyed and the artefacts removed from within the hull. As the excavations progressed the clues as to the identification of the site accumulated, so broader questions could be posed about the nature of the trade the ship and its crew were involved in.

On the wreck site two airlifts were operated within a grid. The airlifting, proceeding from the stern, showed that the vessel lay on its port side, and that the keel and one side of the ship had survived. As the diver operating the airlift exposed artefacts

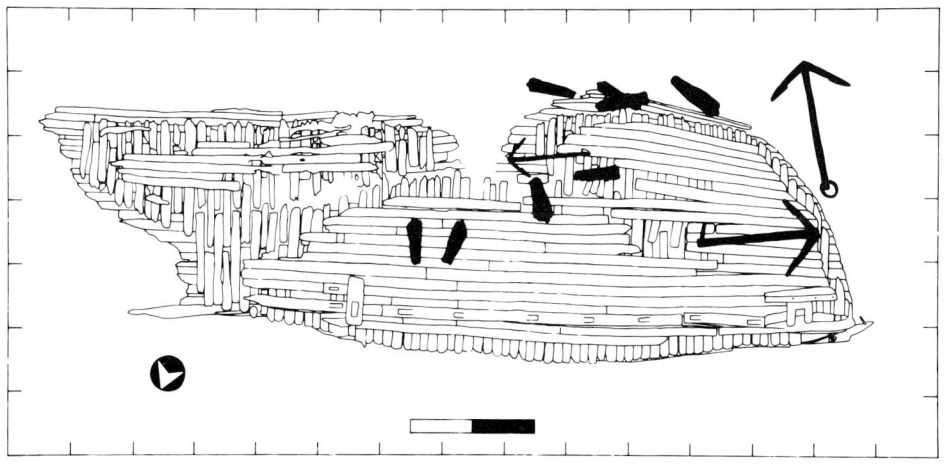

Plan of the *Rapid* site, showing anchors and cannon. (Scale equals 4 metres.)

Divers use a carpenter's level, plumb-bob and tapes to record the profile across the *Rapid* wreck. (Photo: Brian Richards)

the recording diver plotted them on his slate and then put them into plastic bags according to position and type.

The baskets of wet objects brought ashore each afternoon were taken to a sorting table where conservators removed concretions before others commenced the work of registration, drawing, photographing and packing. The more sophisticated conservation treatments were delayed until the items had arrived in the laboratory in Fremantle, but virtually everything from a salt-water environment needs some treatment without delay, so the field conservators worked through the evenings, commencing the chemical, physical and mechanical processes which ultimately clean,

Field conservator Dr James Pang and student Nic Clark prepare parts of a powder canister for packing. (Photo: Graeme Henderson)

Using an air-filled tub, heavy ballast is removed from the *Rapid* site. (Photo: Graeme Henderson)

stabilise, consolidate, conserve and restore the material. The conservator is dealing principally with materials science: it is not really important to him whether a wooden object came from a Viking ship or a twentieth century jetty pile—it is still wood and must be treated as such.[6] So wood was kept in water to which a fungicide was added, iron was kept in caustic soda, and ceramics and glassware were kept in fresh water. Because of the urgent need for dating information the cleaning of coins continued throughout the fieldwork period.

By the end of the first season the after section of the hull had been cleared. Some 18,000 Spanish silver dollars had been found; clearly a consignment for the purchase of cargo. These coins had a date range of 1759 to 1809, with a peak in numbers at 1802–04. The terminal date made it certain that the vessel sank after 1809. The War of 1812, lasting to 1814, reduced trade with the East, so the date of the ship's sinking seemed most likely to have been between 1810 and 1812.

The hull was that of a large, well-armed (eight-gun) ship, which had sailed in ballast. Several hundred tonnes of river-worn stones had to be shifted from the site to expose the timbers, and no cargo was to be seen. It was known that East Indiamen, outward bound from the Atlantic to China or Sumatra, sailed around the Cape of Good Hope and eastward across the southern Indian Ocean. During the unfavourable north-east monsoon, from December to May, these ships took the eastern passage along the north-west coast of Australia and through the eastern islands of Indonesia or, alternatively, sailed around the south and east coasts of Australia, past Sydney. The former route brought numerous ships close to Point Cloates, which otherwise would have been rarely sighted by trading ships.

During the second season of excavation the forward section of the hull was examined. Excavators found an intact barrel of salt beef, marked with the words 'BOSTON MASS. MESS BEEF'.[7] It seemed probable that barrels of provisions for the crew would have been taken on board at the port of origin. In the forecastle area of the hull items were found which would have belonged to the crew, such as a Chinese coin, a Portuguese copper coin and, most importantly, six United States copper one-cent pieces. The one-cent pieces were taken as an indication that at least some of the crew were American, and the Chinese coin seemed to indicate that the ship, or one of its crew, had visited China on an earlier voyage.

So the archaeological evidence pointed towards the vessel being an outward-bound American China-trader or East Indiaman which had sailed from Boston or thereabouts for Canton or possibly Sumatra between 1810 and 1812. Correspondence with American libraries and museums did not provide a name for the vessel, so the Western Australian Museum sent me to Boston to seek an identification. The breakthrough came on the second day in Boston, while skimming the shipping columns of Boston newspapers. The *Columbian Sentinel* for 3rd August 1811 read:

> Ship *Rapid*, Captain Dorr, of and from Boston, has been lost on the coast of New Holland; Captain and crew saved. Captain Dorr and part of the crew of the *Rapid* navigated to Philadelphia the schooner *General Greene*, she having lost her captain and most of her crew at Batavia. The *General Greene* had a

passage of 111 days and arrived last Monday. We learn that almost 200,000 dollars were insured on the *Rapid* in this town. It has been ascertained that the *Rapid* had on board, when she sailed, Two Hundred and Eighty Thousand Dollars in specie. If the ship and cargo have been wholly lost, as reported, it must add greatly to the commercial distress already so severely experienced in this town.[8]

The information in this newspaper article provided the necessary leads to compile a background of the *Rapid* from its building to the time of its loss at Point Cloates on 7th January 1811.

The *Rapid* was built in 1807 by Nathaniel Thomas at Braintree, just south of Boston, for Andrew Ritchie of Boston. The Boston Customhouse Register shows that the 366-ton vessel was built with two decks and three masts, and was 104 feet long, 28 feet 4 inches in breadth and 14 feet 2 inches in depth. She was a square-sterned ship with a figurehead but no galleries. The *Rapid* was registered on 23rd December 1807, just one day after President Jefferson's Embargo Act came into effect. Lacking the naval force to defend neutral rights against England on the high seas, the President had chosen to retreat from the seas altogether. The new Embargo prohibited all American vessels from engaging in foreign trade, and thus began the distressing period of trade disruptions that found their culmination in the War of 1812.

Had the *Rapid* been registered two days earlier, she could have legally left port and commenced a trading voyage overseas. But it appears that the newly completed vessel lay in port unused for at least a year.

She sailed on her first voyage to China in August 1809. The voyage must have been successful because a similar enterprise was undertaken by the owners the following year. The local press announced on 22nd August that specie (coins) would be taken out to be invested in China goods.

The *Rapid* loaded 280,000 Spanish dollars, a large quantity even for those times.

It is likely that she carried no other substantial cargo on her outward voyage. The ballast lying in the hold would have been sufficient to steady the ship under sail. A single barrel of scrap copper, and the remains of a large clock, were the only items found which do not appear to fall into the categories of ship's fittings, provisions, or personal possessions of the crew. These two items may have represented some limited private trade by the ship's master.

Henry Dorr was in command of the *Rapid*. He had been master of the *Jenny* on a voyage to Canton in 1803, and it was he who had taken the *Rapid* out to China on her first voyage in 1809. Henry bought a share in the *Rapid* several weeks before her departure for China in 1810, and owned part of her cargo.

The *Rapid* departed Boston for Canton on 28th September 1810. After rounding the Cape of Good Hope the vessel sailed across the southern Indian Ocean and then north-east towards North West Cape on the Australian coast, where Henry Dorr would have been able to check on his longitude. It looked like being a fast voyage.

But disaster struck at the end of the 98th day. The *Rapid* crashed on to the reef at Point Cloates during the night of 7th January 1811. The next day a storm was

An American ship in distress. (Painting: Francois Roux, 1860)

raging. Finding that efforts to get the ship off the reef were not succeeding, and that she was going to pieces, the crew took to the three boats. Then they set fire to the ship, sacrificing everything so that the wreck would not appear above water and attract other ships to the scene before the captain could return to save the $280,000, all of which remained in the ship.

The wreck was totally on fire when the 22 crew left the area during heavy gales on the 8th. The chief officer with 10 men went in the large boat, the second officer with the carpenter and four sailors in the small one, and sailed for Bali Straits, which both reached safely. Captain Dorr, his clerk Mr Flagg, and three sailors had left in the jolly boat that was a mere 5 metres long and very leaky. The newspaper account of Henry Dorr's boat journey continues:

> After having thrown overboard more than one third of their water, their wet and heavy clothes and other necessary and useful items, the boat's freeboard was still only about one foot. Thus, the voyage was continued towards Bencoolen [a derelict British base near West Sumatra]. Having survived the most imminent dangers of heavy storms and very high seas, they reached Christmas Island [to the South of Sunda Strait] on the 22nd. Here they hoped to find some

water, as they had lived since their departure on three cups of bad water and three small pieces of bread a day. On the 22nd they landed on the NW side and found no fresh water; but then it rained heavily on the next day, and they could catch a considerable amount of water with the sails. They found nothing to eat but beautiful large rats and land crabs, of which they cooked a supply for two days, before setting sail again. On the 28th and 29th they encountered heavy storms from the NW, and hove to. Heavy thunder, strong lightning bolts and screaming winds made their situation most awful. On 5th February they saw Java Point in the ENE, and entered Sunda Straits under continuous lightning toward the Northern Island, where they hoped to find a ship. Disappointed in this they sailed on to Bantam [a port on Java near the northern end of Sunda Strait] arriving on the 8th and being received with the utmost hospitality and kindness, were supplied with provisions for two days. At noon on the 12th they set out for Batavia [approx. 50 miles east of Bantam], but approaching the roadstead at 5 pm the wind died. Fearing that they might be shot at from the fortress, they kept off until the morning of the 13th, when they landed and were offered all assistance from H.E. the Governor General.[9]

Captain Dorr's boat journey had lasted 37 days. All the crew reached Java alive but a number of them died afterwards, including Mr Flagg the clerk. Some 6 weeks after arriving at Batavia the opportunity of a passage home presented itself. The 93-ton schooner *General Greene* had lost her captain, McFarlan, and most of her crew at Batavia, so Henry Dorr and part of the *Rapid*'s crew, presumably his companions from the jolly boat, offered to navigate the schooner to America. They left Batavia during the second week of April and arrived in Philadelphia on 27th July 1811 after a 111-day voyage.

Who salvaged the specie? This is a question which can only be speculated upon at present. After Henry Dorr and part of his crew departed from Batavia on the *General Greene* it would have been simple for any enterprising captain from Batavia or Surabaya to organise an expedition down to the west coast of Australia. If the loyalties of a former crew member from the *Rapid* could be bought, then locating the wreck would be no problem. But even if such a crew member were not available, the contemporary published sailing directions for outward-bound China-traders could be used to narrow the search very considerably. The British hydrographer James Horsburgh in the 1809 edition of his *Directions for Sailing* wrote:

All ships intending to see the coast of New Holland, should make it ... between Shark's Bay and the North West Cape [about 320 kilometres], where there are soundings several leagues from the shore in most places, and it may in general be approached within two or three leagues with safety.[10]

Once in the right area a ship could hug the coast, searching the shoreline with a telescope for the tell-tale concentrations of flotsam which would indicate that the wreck lay nearby. The bow section of the *Rapid* remained intact for at least a year, despite the fire, and at low tides would have been visible for some distance. The wreck lies in a convenient position for salvage, sheltered by the reef from the heavy ocean swells.

Henry Dorr must have contemplated the above scenario at length during his passage home on the *General Greene*. On his arrival in Boston the American ship *Meridian*, which had a cargo of cotton for the East, was obtained and given instructions to visit the wreck, and to claim and recover the property by every means possible. The *Meridian* left Boston on 27th October. A fast vessel on a direct voyage might have been able to reach the West Australian coast by February 1812. But it is possible that some delay could have resulted from a wish to call at an Asian port *en route* to procure divers. Batavia seems likely because Henry Dorr had had ample time to make preliminary arrangements during his earlier stay.

The salvage of some 260,000 coins (the Museum now has the remainder) from the *Rapid* would have taken a number of weeks of good weather. A stable workboat would have to anchor on the poor holding ground near the wreck while divers cut or blasted their way through any structure above the boxes of specie. Then the many boxes would be individually tied and hauled by hand up onto a boat waiting immediately above the wreck. A Spanish dollar weighs about 28 grams, so the consignment of 280,000 would have weighed some 7.84 tonnes. The number of boxes carrying the *Rapid*'s specie is not known, but other examples are available. The Boston ship *Hunter* in 1813 carried $19,400 in 4 boxes—about 136 kg of silver per box, without considering the weight of the box itself. The *Rapid*'s consignment would have filled 58 such boxes.

A salvage party from Java, working covertly on the *Rapid* after Henry Dorr's departure in April 1811 for Philadelphia, would have been acutely aware of the need for haste lest they be caught at the site by a salvage party from Boston representing the owners. On the other hand, the party of salvagers from Boston, operating after February 1812 on the wreck, would have been aware of the deteriorating relations with Britain and the danger of being captured by a privateer while at Point Cloates salvaging the treasure. These pressures may explain why some $19,000 were left on the site. In May 1812 the *Meridian* arrived at Java from the wreck with the news that they had salvaged very little. Captain Wooddang brought the misleading information that Britain and America had settled their differences.

It seems likely that Captain Wooddang then recruited Javanese divers and returned for more successful work on the wreck, because in February 1813 the Canton agents for the *Rapid* announced 'the arrival of ship *Meridian* at Canton with $90,808.25 of specie belonging to the proprietors of the money lost in ship *Rapid*'.

Henry Dorr and other members of his family suffered financially from the wreck: Henry lost $9,000, his brother Samuel nearly the same, and the estate of his father $2,553. But there is no indication that Henry was ruined. He may have come out to the East with the *Meridian* during the years 1812–15. In June 1816 he left Boston for Canton as master and supercargo of the *Ida*. In 1819, with the same vessel, he utilised the favourable monsoons to make a fast passage of 101 days from Canton to Boston. He died in 1850 of jaundice, aged 70.

Although a broad background to the *Rapid* has emerged from the fieldwork and archival investigations, a number of basic questions still await answers. For example, what were the circumstances that led to the ship's loss? Did the night watch fall asleep or was there a faulty chronometer on board? Were the gales referred to part of one of the tropical cyclones which strike the coast during the summer months?

The available details regarding salvage are even more sketchy. Did the salvors establish a shore camp? Did they rely entirely on their own water, or dig wells on the shore? Did they meet the local Aborigines? The answers to these questions will increase our knowledge of foreign contact with Western Australia prior to the first settlement in 1829.

Other questions, equally important, relate to the development of international trade, and to shipbuilding in America. The *Rapid* is the first example of an outward-bound American China-trader to be given archaeological attention, so aspects of that trade can now be re-examined from the archaeological standpoint. There are large gaps remaining in the history of shipbuilding developments in America during the important post-revolutionary period, and the hull of the *Rapid* provides the opportunity to fill some of these gaps.

3. The James Matthews Shipwreck Excavation

The identity of the *James Matthews* wreck was known before the excavation commenced. The early background of the vessel emerged after work had begun, and influenced the aims of the later excavations and the methods employed.

The 25-metre snow brig *James Matthews* was probably built in France early in the nineteenth century.[11] In 1836 a Frenchman, Gabriel Giron, sold her and the vessel assumed Portuguese registration. She was then named *Don Francisco* after her new owner, the notorious Don Francisco Felis da Souza, alias Char Char, slave dealer of Whydah. Da Souza was a close friend of King Ghezo of Dahomey and a one-time governor of the Portuguese fort of Whydah on the Slave Coast in the Bight of Benin. He owned a fleet of ships. His friend King Ghezo was the most feared man in Africa because of his regular human sacrifices. In the words of an old ditty of the time:

> Beware and take care of the Bight of Benin
> There's one comes out for forty goes in.

Of those leaving, the overwhelming majority were destined to eke out their days as slaves on the Caribbean plantations.

Early in 1837 the 107-ton *Don Francisco*, crammed full with 433 slaves, crept out of Whydah on an illegal voyage to Havana. She was commanded by Captain Antonio Pieria Lisboa, and carried a crew of 34 as well as four passengers. Captain Lisboa successfully eluded the British naval patrols watching the African coast but ran into trouble near the conclusion of his voyage. On 25th April, 1837, after a 7-hour chase, his vessel was seized as a prize by HM Brigantine *Griffon* off the island of Dominica, one of the windward islands of the Lesser Antilles. The *Don Francisco* was found to be in a near-sinking state and was taken in to Dominica instead of making the normally required long voyage back to Freetown, Sierra Leone, for adjudication. She was condemned as a slave trader by the British and Portuguese

The chess set which led to a fleeting shipboard romance between a passenger on the *James Matthews* and another from a vessel sailing in company. The pieces were found on the wreck site. (Photo: Pat Baker)

Mixed Commission Court on 21st November 1837, but was subsequently reregistered and named *James Matthews*. The vessel was then entered into legitimate trading activities which eventually took her to London.

The *James Matthews* left London for Western Australia in March 1841 with 7,000 slate roofing tiles, farm implements, general cargo, three passengers and a crew of 15. Henry de Burgh, a young passenger, left a diary covering the voyage to Australia. De Burgh owned part of the cargo and had a financial interest in the vessel itself.

After arriving safely at Fremantle in July the *James Matthews* was moored offshore to commence discharging cargo. But a day after her arrival a series of heavy squalls struck the port and the vessel was blown ashore on the north side of Woodman Point. The masts were cut down but one pierced the bottom of the brig and it soon filled with water. Henry de Burgh lost a case of guns and rifles, and a chest containing 200 sovereigns.

Skindivers rediscovered the wreck on 22nd July 1973, 132 years to the day after the sinking. The wreck lies 200 metres from shore in 3 metres of water on a sandy bottom. The highest section of the wreck consisted of a mound of slates and bundles of iron rods, comprising paying ballast which had been destined for colonial blacksmiths, who in those days relied entirely on imported supplies of materials. Deeper down were the waterwashed ballast stones which covered the ceiling planking of the hull. Proximity of the site to an industrial complex sometimes affected visibility: a milky suspension cloud of lime regularly moved over the site, reducing visibility to less than one arm's length. But the wreck was rarely affected by heavy seas so there was no need for sacrifice of archaeological method.

The first stage of the excavation consisted of airlifting to remove sand which covered the site by up to 1.5 metres in depth. We had to remove a large quantity of sand (180 cubic metres) at least 3 metres to either side of the wreck in order to completely expose the site and avoid substantial re-fill during the excavation season. The shallowness of the site presented some problems in terms of the airlift, that underwater vacuum cleaner which is the maritime archaeologists' standard tool for clearing sand from a site. The airlift relies on the fact that air released into a tube on the seabed will rush up that tube to the surface, producing a vacuum effect on the seabed. In deeper water the pressure increases, improving the upward flow of air released into the airlift tube. But on shallow sites the efficiency decreases, so on the *James Matthews* site a large 3.5-cubic-metres-per-minute compressor was used to ensure a large volume of air for the airlift.

After the sand had been removed from the wreck, grids were set up so that archaeologists could draw the wooden hull of the ship. The vessel lay on its starboard side. The deck and the port side had long since disintegrated but the full length of the keelson had survived, together with the structurally intact starboard side of the vessel. A 30 by 6 metre (180 square metre) rectangle was marked out around the structure and a mobile 1 by 6 metre grid, shaped like a bed frame, with four legs, two sides and two ends, was erected across one end of the site. Then the height and level were adjusted to a horizontal position using a carpenter's level and an underwater plastic bubble-tube. When the area under the first grid position had been recorded, sleeve screws at each end of the grid were loosened, and the end bars slid horizontally forward to take up the second grid position. For recording, a slide similar to that on a standard slide rule was placed on rails so as to run from one end to the other of the grid frame. A plumb-bob was hung from the slide to give the positions of the objects being recorded.

Divers drawing underwater used pencils on sheets of waterproof drafting film taped to perspex boards. A separate sheet was used for each of 30 grid positions. First the diver drew a sketch plan of the material lying below his grid and numbered each significant point. Then, for each of the numbered points, three co-ordinates were measured. The X co-ordinate was the horizontal distance along, the Y co-ordinate was the horizontal distance across, and the Z co-ordinate was the vertical distance below the grid plane. This survey technique gave accurate and detailed three-dimensional data about the wreck.

Divers discard fins in order to measure the hull and slate cargo of the *James Matthews*. (Photo: Pat Baker)

On land, transfer of the survey data onto graph paper was a straightforward but lengthy procedure. A detail plan of the wreck site from the X and Y co-ordinates has been prepared at a scale of 1:10, and the X and Z co-ordinates were used to draw profiles of the ship at regular intervals. The next stage will be to draw the lines of the incomplete vessel. It will then be possible to make both general and particular observations about the construction of the vessel in terms which have meaning to naval architects.

Photography was an important part of the hull survey. A tubular-steel pyramid, with two cameras mounted side by side at the apex, was positioned on the grid. Then a series of overlapping stereo photographic pairs were taken of each grid area. The stereo pairs could be viewed under a mirror stereoscope, giving an enhanced three-dimensional view of site details and relief. They were also used to make up photomosaics of each grid area. In these ways the photographic record was an aid to the interpretation of complex details on the site when the plan was being drawn up.[12]

The vessels built for the slave trade during the nineteenth century, when it was illegal, needed to meet special constructional requirements, such as a shallow draft for hiding among the mangroves on the African coast, fine lines for speed to escape apprehending pursuers, and various internal fittings to provide for the slaves. Slave ships were famous for their excellent sailing qualities and some of their internal fittings were so obviously intended for the housing of slaves that legislation was passed providing for vessels of this design to be detained as prizes even when no slaves were found on board. Existing plans of slave ships during the illegal period are scarce and give limited information, so surviving hulls are the only potential source of detailed information on this type of vessel.

The *James Matthews* hull is particularly important because it is the only surviving representative of the Atlantic slave trade yet located. The internal fittings were doubtless altered when the vessel was re-fitted for legitimate trade, but the hull has been well preserved and the underwater environment is favourable for detailed recording and delicate excavation. One wreck does not provide sufficient data on which to base sound historical or anthropological theories. A number of wrecks from the same trade must be excavated, interpreted and published before such data, used together with conventional documentation, can form the basis for such theories.

In time, other examples will doubtless be found, and these will, on detailed examination, provide a comparison with the *Don Francisco* hull. The 91-ton brig *Schah*, a slaver, was condemned at Sierra Leone in 1832, and later, as a legitimate trader, was wrecked on the Gippsland coast of Victoria in 1837. If her remains prove to be substantial when found she will provide such a comparison.

The *James Matthews* ended its career not as a slaver, but as a general trader bringing settlers to Western Australia, and this aspect can be illustrated by the cargo. During four seasons of excavation, roofing slates were the most bulky commodity raised. About 3,500 were found unbroken, but the remainder had been shattered when the vessel sank. Other farmhouse construction materials included several cases of glass window panes and dozens of heavy door hinges. Carpenters' tools included awls, parts of saws, a wood plane, wooden tool handles, and small nails. Among the domestic items were a large number of smoothing irons chained together in batches, birch brooms, stoneware, tobacco and preserved fruit jars, china dishes, glass tumblers, bottles of wine, a candlestick, an umbrella, shoes, part of a chair, a number of heavy cast-iron cooking pots, and about 500 clay pipes. An almost-complete chess set was possibly the possession of Henry de Burgh, who reveals in his diary that he developed a fleeting shipboard romance over a chess-board tussle with a young lady visiting from a passing ship.

4. Other Colonial Period Fieldwork

Full-scale excavations represent a relatively small part of the work carried out in the field by the maritime archaeology department and its associates. Other activities include the ongoing inspection programme, Museum and contracted pre-disturbance surveys, environmental impact reports, salvage archaeology, test excavations, and attention given to maritime sites found on land. Members of the Maritime Archaeology Association of Western Australia make their contribution to a number of these activities.

Wreck inspection generally proceeds along the following lines. A diver visits the Museum and reports finding artefacts on the seabed. Museum staff make a cursory examination of any artefacts brought in, to decide whether it is likely that a shipwreck site has been found. If this does appear so then the finder is asked to point out the locality of his site on one of the Museum's series of sea charts. These charts are

Archaeologist Scott Sledge checks corrosion potentials on a pintle lying on the Rowley Shoals shipwreck site. (Photo: Pat Baker)

already marked with the locations of shipping casualties as indicated by archival records. If the nature and origin of the artefacts appear to correlate with any of these vessel names then the sites' register is consulted to obtain a background of those vessels. Museum staff then arrange to visit the site with the finder. At the site the wreck's position is recorded by means of sailing directions, compass bearings, visual transits (photographs are taken), and sextant angles. The biophysical environment is recorded and the wreck itself measured and described. Samples are frequently taken from the ship's structure or cargo, for future analysis.

Back at the Museum, the site location is recorded in terms of latitude and longitude co-ordinates. The sample artefacts and site description are again compared with archival sources about known casualties in the area with a view to identifying the site. After due consideration the Museum's Maritime Archaeology Advisory Committee sends information to the Director about the significance of the site as indicated by standard guidelines, and makes recommendations as to whether the site should be protected under either of the Acts.

Where site security is not jeopardised, the Museum endeavours to increase the accessibility of underwater sites to recreational divers and glass-bottomed-boat viewers. On Rottnest Island, close to Fremantle, major wreck sites are marked on the seabed by concrete obelisks bearing interpretive data. Similar information is available on the shore adjacent to each site, for the interest of non-divers. When the historic steam-powered whale-chaser *Cheynes 3* was due to be scuttled in deep water off Albany, local divers approached port authorities in a successful bid to have the wreck sunk in sheltered waters shallow enough for scuba divers to experience the exhilaration of wreck exploration. The nearby Whale World Museum has more information about the whale-chasers. Such underwater interpretive projects increase public appreciation and understanding of the cultural heritage without leading to the destruction of the resource.

Pre-disturbance surveys are generally conducted using photogrammetric techniques or trilateration from a base line. Some sites are exposed or made accessible only by temporary seasonal conditions, and in these circumstances local diving clubs or interested individuals are sometimes contracted to conduct surveys. The most useful of these was the pre-excavation survey of the *Vergulde Draeck* wreck, used as the basis for all the Museum's excavation work on that site.

Before any excavation work commences on a wreck site the Museum's conservators monitor the rate of corrosion by obtaining electrochemical data. The data obtained by the conservators is frequently of application both inside and outside the field of maritime archaeology. By comparing sites they have shown for example that the corrosion mechanisms for copper and its alloys are dependent on site location and the formation of calcareous concretions, which establishes a microenvironment significantly different to ambient sea water.[13]

Environmental impact reports and salvage or 'rescue' archaeology have been conducted under a variety of circumstances with variable results.[14] The State legislation does not yet protect sites later than the year 1900, and initially did not protect sites above the high water mark. At Bunbury in 1973 Museum staff were given just several days to examine a large wooden shipwreck unearthed by the Public Works Department several hundred metres inland, before bull-dozers reduced the hull to matchwood. Similar circumstances applied when a dredge, commencing deepening operations at Careening Bay, Garden Island, came upon an obstructing hulk in 1973. The Maritime Archaeology Act of 1973 broadened the legislative protection given to shipwrecks, and subsequent port developments have generally taken ship-

The fate of a shipwreck site not protected by legislation in 1973. (Photo: Lous Zuiderbaan)

wreck sites into account. So when continuing dredging operations at Careening Bay exposed a second hulk the Commonwealth Department of Works arranged for it to be shifted into deeper water where it could be examined at leisure and re-buried. The hulk proved to be that of the barque *Day Dawn*, built as a whaler in Massachusetts in 1851.[15] Similarly, in 1978, when plans were announced to expand the shipbuilding industry in Jervoise Bay, Cockburn Sound, the State Department of Conservation and Environment provided funds for the sites in the area to be located and evaluated. A less desirable set of circumstances occurred in July 1984 when the Museum was given several weeks' notice that rock fill was about to commence for a marina impinging upon a section of the historic long jetty which formed Fremantle's first deep-water harbour off Arthur's Head during the nineteenth century.

Limited excavation work has been carried out on a number of sites for special purposes. Subsequent to the 1981 discovery of the wreck of the 444-ton British-India ship *Cumberland* (lost near Cape Leeuwin in 1830 and representing the earliest post West Australian settlement shipwreck site yet found), reports came to the Museum that exposed material on the site was being removed illegally by souvenir hunters. Museum staff, assisted by local divers and members of the Maritime Archaeology Association, spent a week surveying the site and removing exposed artefacts considered to be in jeopardy. The largest artefact raised was a 3-metres-long 18-pounder

Using air balloons, a cannon is lifted from the shallow *Cumberland* (1830) wreck.
(Photo: Jon Carpenter)

iron cannon, bearing the symbol of the Dutch East India Company, and carried as part of the ship's ballast.[16]

More substantial excavation took place on the cargo remaining on the wreck of the barque *Eglinton*, wrecked to the north of Fremantle in 1852. The vessel had been carrying 30 British emigrants and a heavy general cargo for the colonial storekeepers and the Commissariat. Very little of the structure of the hull has survived but casks, cases and packages of assorted goods, thrown overboard with the intention that they float ashore, had sunk, and found their way into small protected tunnels which honeycomb the limestone reef. The excavation of these tunnels was interesting because of the cargo items yielded. The *Eglinton* had been carrying the first general cargo intended for the neglected Swan River Colony for many months so the items on the wreck can be expected to reflect the community's annual import needs. There were no manufacturing industries in the Colony: part of its *raison d'etre* was to provide a market for Britain's burgeoning industrial output. There were bricks for building houses, weights and scales, mortars and pestles, smoothing irons, candlesticks, taps, parts of lamps, medicine bottles, wine bottles, glass decanters, wine goblets, dinner sets, a pewter teapot, a comb, salt cellars, and a variety of preserves containers. Most of these items were intended for the storekeepers' shelves, and would then have gone to settlers' cottages. The glassware and ceramics were of a cheap, durable quality, with a utilitarian rather than decorative function. Many of the dinner sets were of ironstone body, the most common being a Minton underglaze transfer pattern called 'Anemone'. The hundreds of roughly moulded glass tumblers may have been on their way to a working-men's ale house. Water filters were in everyday use because the struggling colony had not yet developed suitable water supplies. Several were found on the wreck.

The London Times announces the departure of 'the splendid fast sailing ship *Eglinton*' on what was to be her last voyage.

Part of the collection relates to the convict establishment. The Colony had begun to receive transported convicts from Britain in 1850 with the hopes of injecting capital and free labour into the flagging economy. The *Eglinton* carried a variety of goods for the convict establishment. Items raised include shoes marked with the broad arrow of the convict system, musket balls, percussion caps, pick and axe heads. Signs of the inevitable growth of social awareness in the Colony are to be seen in copper printing plates for the printing of introductory calling cards. Examined as a whole, the collection reflects a small, isolated and underdeveloped society tied firmly to the distant mother country.

When the loss of the *Eglinton* was reported in Fremantle it represented a threat to the well-being of the entire Colony because £15,000 in bullion, intended for the Government, was trapped in the wreck.

A schooner was chartered to go to the wreck. On board, with instructions to salvage the bullion for the Government, were George Clifton (the Superintendent of Water Police), and a ticket-of-leave convict named Rodriguez who was a skilled diver. But the insurers had announced that the bullion was worth £5,000 in salvage money, and this tempted the enterprising rascal who was skipper of the schooner. On arrival at the reef, Clifton rowed across to the wreck to make arrangements for the salvage job, and the skipper, seizing the opportunity, weighed anchor and made sail, with the diving equipment still aboard. All he had to do was stand off until Clifton tired of waiting and rowed ashore. Then he would return to the wreck, salvage the gold and claim the salvage money.

But Clifton was not to be put off. He rowed after the schooner, overtook it, and fired two shots close to the helmsman, forcing him to stop. Then Rodriguez went down into the wreck and successfully recovered the bullion from the hold. As each chest reached the deck the Superintendent insisted on personally slinging and lowering it into the police boat. Although a civil servant, Clifton was not one to let opportunity go by. Later he laid a salvage claim to the Government on his own behalf.

The wreck of the screw steamer *Xantho*, lost north of Geraldton in 1872, has been the subject of an intensive survey. The iron hull of the *Xantho* was built in 1848, quite early for a steamer. But it is the steam engine—a 60-horsepower horizontal trunk compound engine built by Penn and Son of London—that is currently of most interest to the Museum, for exhibition and research purposes. The engine appears to the chemists and corrosion scientists, who examined it underwater, to be in excellent condition. Museum staff recently raised the engine, and it is now being restored to place on display.[17]

Maritime archaeologists have occasionally ventured ashore to join other Museum staff in examining dry land maritime archaeological sites. During the past decade Museum staff and contracted archaeologists have conducted surveys of the eighteenth century *Zeewijk* wreck survivors' camp site on the Abrolhos Islands, early maritime explorers' camp sites in the North West, nineteenth century pearling and bay whaling sites in the North West, an early twentieth century shore-based whaling station at Norwegian Bay, and twentieth century guano loading and mining facilities on the Abrolhos Islands.[18] More recently, contract archaeologists working on a

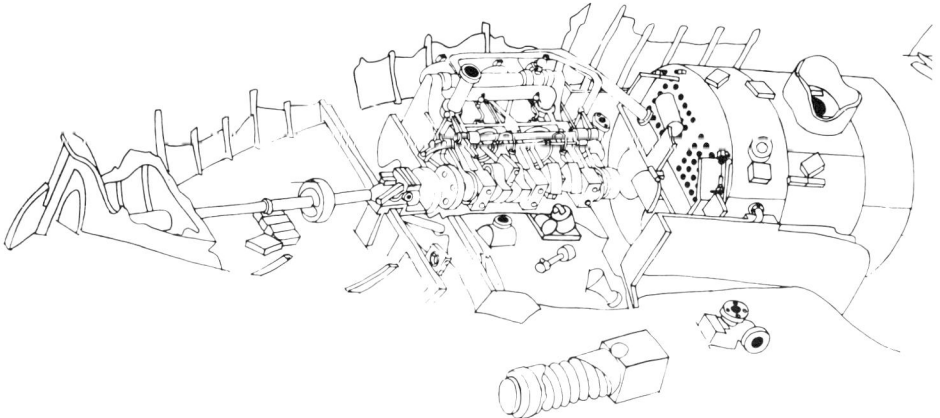

An isometric sketch of the SS *Xantho* wreck shows sternpost, steam engine and boiler. (Drawing: John Riley)

Conservators Ian MacLeod and Nancy Mills-Reid join archaeologist Mike McCarthy in de-concreting the steam engine from the wreck of the SS *Xantho*. (Photo: *West Advertiser*)

Cheynes 3, a whale-chaser, is sunk with explosives in shallow water in 1983 to make the wreck accessible for recreational divers to explore. (Photo: Graeme Henderson)

A diver swims along the deck of the scuttled whale-chaser. (Photo: Pat Baker)

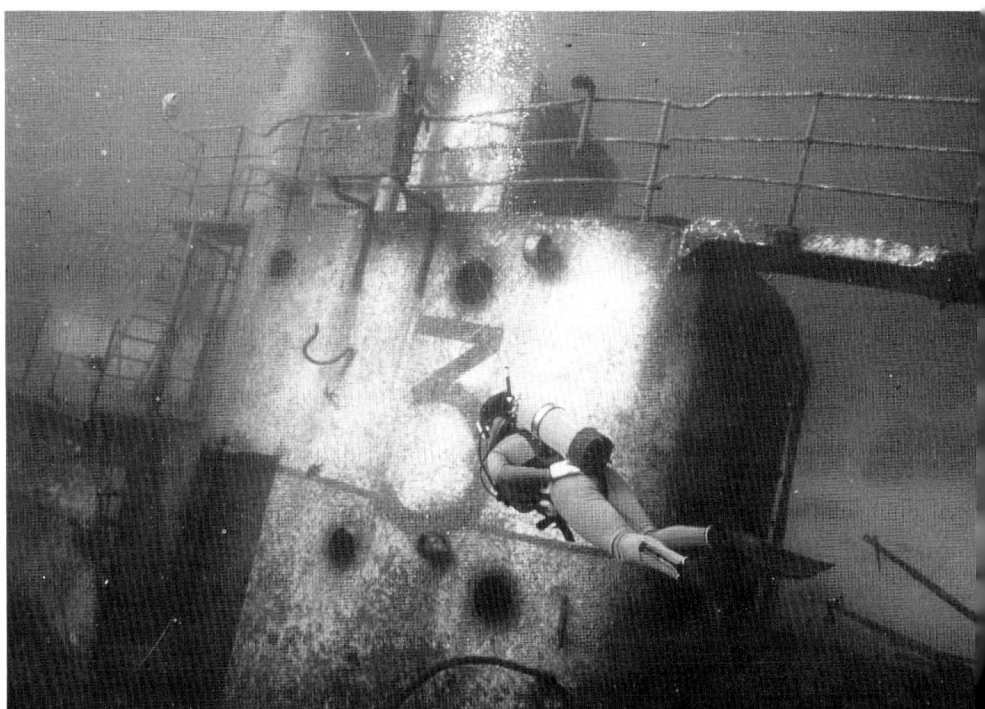

National Estate grant opened several test trenches to successfully locate the position of early nineteenth century bay whaling tryworks and boatbuilding works in Bathers Bay, Fremantle. The bay whaling enterprise of the 1830s was the infant Swan River Colony's first industry so the well-preserved tryworks are of great public significance.[19]

The Maritime Archaeology Association of Western Australia was formed in 1974 to bring together amateurs interested in the work being done by the Maritime Archaeology Department. It developed three distinct roles during the past decade. Firstly, it has been the source of much of the voluntary labour and expertise used on Museum fieldwork projects. Association members have a broad range of experience and qualifications, and include surveyors, engineers, doctors, lawyers, professional divers, mechanics, and many other skills of practical value to the Museum's limited staff numbers in the field. Secondly, Association members organise their own projects, utilising experience gained on Museum fieldwork projects. The emphasis is towards non-disturbance activities and includes archive searches for historical backgrounds to particular sites or groups of sites, searches underwater to locate previously unfound sites in areas where industrial developments might lead to their destruction, site surveys utilising various methods, and experimentation to improve field equipment.[20] The third, and perhaps less conscious function of the Association, is to provide an interface with the general diving public. Many members stay with the Association for a relatively short period of time, their chief love being diving rather than archaeology. But when they leave they generally take with them an enhanced awareness of the Museum's objectives and of the significance of shipwreck sites as cultural resources, and they share this awareness with other divers.

Chapter 7 REFERENCES

1. Henderson, 1980 (3).
2. ICOMOS, 1981.
3. McCarthy, 1979.
4. Sledge, 1977.
5. Henderson, 1983, pp. 3-4.
6. Pearson, 1977, p. 116.
7. Henderson, 1981 (2), p. 125.
8. *Columbian Sentinel*, 1811.
9. *Bataviasche Koloniale Courant*, 1811.
10. Horsburgh, 1809, p. 96.
11. De Burgh and Henderson, 1979, p. 9.
12. Baker and Henderson, 1979, pp. 225-244.
13. MacLeod, 1984, pp. 1-10.
14. Rahtz, 1974.
15. McCarthy, 1979, p. 157.
16. Sledge, 1984, p. 3.
17. McCarthy, 1984, p. 5.
18. Ingleman-Sundberg, 1978; MacIlroy, 1979; Stanbury, 1982.
19. MacIlroy and Meredith, 1985.
20. McCarthy, 1979 (2).

8 Adventure and Misadventure in the South Pacific: Maritime Archaeology at the Queensland Museum

No State legislation exists in Queensland for the protection of historic shipwrecks. Maritime archaeology was initiated in Queensland because of the presence of one shipwreck in particular—HMS *Pandora*.

The *Pandora* was discovered in November 1977. Both the Commonwealth and the State Governments were agreed that the *Pandora* site, once discovered, should be given protection, so the Commonwealth Historic Shipwrecks Act was immediately proclaimed to apply to waters off the Queensland coast. Later, as an added protective measure, the site was made a Protected Zone, and was given the benefit of surveillance by the Australian Coastal Surveillance Organisation. The finders were paid a reward of $10,000 by the Commonwealth Department of Arts, Heritage and Environment. Shortly after the implementation of the Act a committee, representing the Department of Harbours and Marine, the Queensland Museum, and the Queensland Maritime Museum Association, looked at the question of which State body should undertake maritime archaeology.[1]

The Queensland Maritime Museum Association has a large block of land on the bank of the Brisbane River, and its facilities include an indoor maritime museum, a dry dock containing the de-commissioned HMAS *Diamantina* and several large workshop-storage sheds. The Maritime Museum, a private association, is made up of an enthusiastic group of volunteers who have brought together and now curate the collections and who open the Museum to the public.

The Queensland Museum has a broad range of subject interests and is funded by the Queensland Government as a part of the Public Service Board. Like the Western Australian Museum, it has a Board of Trustees, but the Queensland Museum, with a staff numbering just over 100, is somewhat smaller, and its responsibilities do not penetrate to the regional centres. It now has a Department of Maritime Archaeology (permanent staff of one) and a small Department of Conservation which serves the entire Museum.

The Queensland Museum was recommended as the competent authority for maritime archaeology and the delegation of responsibility was formalised in July 1981. The Curator, Ron Coleman, was appointed in 1982. Since that time, planning for the excavation of the *Pandora* has been the central thrust, but work has also commenced on a sites' register, and a number of significant shipwreck sites have been inspected. Liaison with diving clubs, in particular the Underwater Research Group, has encouraged the development of an active Maritime Archaeology Association of Queensland.

1. The Excavation of HMS Pandora

The continuing excavation of the *Pandora* site is surely the most exciting current maritime archaeological fieldwork being conducted in Australia. The wreck was discovered by divers from two private vessels operated by Steve Domm and Ben Cropp. The vessels' skippers were competing for the honour of being first to locate the wreck. They were searching in the vicinity of Pandora Entrance, with the assistance of an RAAF Neptune aircraft carrying a magnetometer. This instrument measures terrestrial magnetism, and the searchers hoped that it would be sentitive enough to register the iron cannon, anchors and ballast of the *Pandora* if the Neptune's flight path took it over the wreck site. The Neptune crew reported a reading close to a reef in 30 metres' depth of water. This was checked and it proved to be a wreck site littered with cannon and large anchors.

After ensuring that the site had the full protection of the law, the Commonwealth Government wanted to be convinced that the wreck found was indeed the *Pandora*. In January 1979 the then Prime Minister, Malcolm Fraser, sought the assistance of the Western Australian Museum for an assessment of the site. The Department of Arts, Heritage and Environment arranged for an expedition, led by myself, to identify the wreck, assess its significance as an archaeological site, and comment on the possibility of future excavation of material for research and display.

In April 1979 the expedition members assembled at Thursday Island, an old pearling port off the tip of Cape York Peninsula—the northern extremity of the Australian continent. Pat Baker of the Western Australian Museum was my photographer and diving safety officer, while Ben Cropp and Steve Domm acted as guides. The 27-metre Department of Transport vessel *Lumen*, which was ready for a voyage to Cairns, was diverted to Pandora Entrance, about 200 kilometres E.S.E. of Thursday Island, so that we could examine the site.

Lumen arrived at Pandora Entrance at 2 p.m. on 20th April 1979. The vessel was manoeuvred to roughly accord with the compass bearings which the discoverers had taken to three tiny sand cays lying on the horizon. The echo sounder showed a very slight irregularity on the seabed, and a marker buoy was thrown overboard. At 3.05 p.m. we entered the water and, somewhat to my surprise, immediately located the wreck in 30 metres of water.

Visibility was exceptionally good: the entire length of the 40-metre site could be seen by a diver swimming 15 metres above the seabed. The wreck lay in a gently sloping bed of coarse white coral sand on a 150°–330° magnetic axis. There were no rock outcrops and coral growth on the wreckage was minimal.

The tight distribution of wreckage indicated that the ship had not broken up as a result of any underwater turbulence. Rather, it had settled into the seabed and the marine worms gradually devoured the wood construction of the upper works, allowing iron cannon and other heavy durable objects to sink vertically into the

Turtles lay their eggs on a group of sand cays, one of which was visited briefly by the survivors from the *Pandora* wreck. (Photo: Pat Baker)

sand below. Exposed copper sheathing at one end of the site was from the sternpost. No wood was visible above the sand; marine organisms would have long since digested any exposed timbers. But the conditions on the site—deep water, no turbulence on the seabed, a moderate tidal current, level deep sand around and over the wreckage—favoured the survival of a very substantial portion of the wooden structure buried in the sand. Exposed sections of chain plates (which secure rigging to the sides of a sailing ship's hull) gave some indication of the level of timber destroyed on one side of the ship.

Swimming from end to end of the site, the major items we saw were as follows: a large anchor with one arm erect, a single small area of moulded copper sheathing standing proud of the bottom (which seemed to indicate that the teredo worm reached the vessel's waterline only at this section, because the entire hull below the waterline would be expected to have originally been clad in copper sheathing), copper tubing (which proved to have been the vent from a cabin fireplace), a group of iron cannon, a drive wheel from a chain pump, the ship's iron stove, several large 'Ali Baba'-type oil jars, more anchors and bronze rudder fittings. Lead deck scupper pipes, chain plates and a variety of other iron work were intermingled with this material. The copper and bronze fittings were in good condition and the large ceramic jars were intact. Iron objects such as cannon and anchors had retained their shape well, and were only lightly encrusted.

The first dive lasted 25 minutes. Time was the enemy in this deep water. We remembered that in 1977 a member of the discovery team had overstayed his bottom time, and suffered an attack of the bends as a result.

Next day, during two hectic dives totalling 45 minutes, divers set up a survey base line, trilaterated part of the site, obtained photomosaic coverage for the full site and briefly examined the principal exposed features. But then the weather deteriorated overnight, preventing any prospects for further diving, not because of any underwater turbulence, but because the rolling motion of the *Lumen* in the sloppy sea conditions prevented the launching of the aluminium dinghy used as a diving tender. So the *Lumen* began its 5 days' voyage to Cairns, expedition members having spent just over an hour each on the seabed.

With the site inspection accomplished, the immediate aim was to identify the wreck. The first thing to be considered was that Cropp and Domm had found the site after a planned search. They had studied the accounts of the *Pandora*'s survivors and these had led them to within a kilometre of the wreck. Pandora Entrance with its sandy cays can be seen on Admiralty charts of the area. Captain Edwards gave the depth of water, the distance and the bearing—West by South half South—from the sinking *Pandora* to the islet on which he landed. The searchers had to compensate for changes in magnetic variation since 1791 and guess which of the three cays in the group Edwards was referring to. The magnetometer was an important search aid, but only after the location had been narrowed down. Yet it could not be assumed simply because this wreck was in the correct position that it was necessarily the *Pandora*. The real test of identity had to be a study of some of the artefacts from the wreck.

A small number of items had been raised from the site by the finders, and these, on close analysis, provided a lot of information about the origins of the wreck. The salvaged items included two bronze rudder pintles, one of which was cleaned and treated at the Western Australian Museum's Department of Materials Conservation, revealing a number of significant markings. Two embossed broad arrow marks showed that the pintle was made for the British Government, and indicated that the wreck was that of a British naval vessel.[2] A series of small dots had been punched into the piece to form the number '24'. The number of guns was the most precise way for those in the shipbuilding industry to refer to a particular class of vessel. *Pandora* was a 24-gun, 6th rate, and the number on the pintle referred to this class. The name Forbes also appeared in embossed lettering. It undoubtedly referred to William Forbes, who in 1779 was supplying a large proportion of the copper items used at the Deptford Yard, where the *Pandora* was built. The British Navy's standard building contract in 1782 required that 24-gun ships were to have five pintles on the rudder. This same contract states that the pintles were to be 2-3/8" (0.06 metre) diameter, the same as the two pintles raised from the wreck site. So this one object showed that the wreck was that of a 24-gun British naval vessel built at Deptford. Other items raised offered confirmation that the wreck was the *Pandora*. One example was the spectacle plate.

This rudder pintle, cleaned in the Western Australian Museum's conservation laboratory, bears a broad arrow, the name 'Forbes', and the number 24—information which helped greatly in the identification of HMS *Pandora*. (Photo: Pat Baker)

The building contract stipulated 'a ring bolt with two rings by one inches (0.025 metre) diameter drove through the rother (rudder)'. This fixture with its rings or spectacles was a feature of vessels built for action. When a rudder sustained action damage it was generally below the tiller. A special bronze spectacle plate consisting of a band containing two spectacle extensions with eye holes was therefore provided above the heel of the rudder. Ropes were rigged and shackled to the spectacle plate eyes. These ropes were led inboard through shackles secured to the hull and to manned tackles inboard, thus improvising a simple and effective secondary steering arrangement. When the *Pandora* struck the shallow reef the sternpost sustained damage, resulting in sections of the rudder being broken away. When the wreck site was discovered, divers sought verification of the wreck's identity by searching the adjacent reef top for rudder fittings, and found the spectacle plate, which has an eye diamater matching that specified in the building contract, and is stamped with the name 'Forbes'. The presence of the spectacle plate was fully consistent with a British warship of the late eighteenth century, and its position on the reef top matched the circumstances of the *Pandora*'s loss.

The other items raised from the site at that time—part of a pump, a large oil jar, a copper fastening bolt, a small salt-glazed stoneware jar, a glass beverage bottle and several bases, and several small fragments of bone and wood—were all fully consistent with the *Pandora*.

The pump section raised consisted of heavy-gauge copper tubing from the lower barrel of a common or hand pump. The lower end of the tube has a shoulder and an everted lip to facilitate the connection with the pump nozzle, which was usually of lead. The most useful pump was the chain pump (producing a continuous flow of water) which was universally used in the navy by that date. But Captain Edwards's log shows that the *Pandora* carried both common and chain pumps. On the night of the ship's loss he wrote:

> Soon after the ship was over the reef one of the chain pumps gave way by the chains breaking and a little before midnight one of [the] hand pumps was rendered useless for some time by the spear box breaking.[3]

The oil jar raised by the finders from the bow section of the wreck was made of pink earthenware and showed no sign of glazing. It was 0.75 metres high, with a thickened rim around the mouth and two vestigial arched handles enclosing an applied plaque with the raised letters 'TMF' under a Lorraine cross. Similar jars of west Mediterranean origin have been found in Britain and the United States, and have been dated to the second half of the eighteenth century.

The evidence seems to point to Tuscany in Northern Italy as the place of manufacture of this sort of jar: Tuscany belonged to the House of Lorraine between 1737-1859, and later in the nineteenth century many of the jars bore Tuscan potters' marks.[4] They were generally used as oil jars but books on eighteenth century shipbuilding show similar jars employed on ships and described as water beakers.[5]

The peculiar nature of the *Pandora*'s mission required a large quantity of fresh water to be readily available in the vessel for the extended crew and the bread-fruit plants, although presumably it would have been transported out in barrels. The *Pandora* is known to have called on its outward voyage at Santa Cruz, Teneriffe, for water and wine, and at Rio de Janeiro to complete her water. Santa Cruz seems a likely place to have picked up Italian jars. However, they are known to have been available in Britain and may have been supplied before the *Pandora* commenced her voyage. Among the alterations made to the *Pandora* in fitting out for the voyage was:

> A small place to be berthed in with thin slats, forward between decks for the stowage of oil jars, and such irregular formed things, as would occasion considerable breakage if stored in the hold.[6]

Considering all the available evidence, it was clear that the site investigated was the *Pandora*. The wreck lay at Pandora Entrance, close to breaking reefs, in the depth of water indicated by the survivors. No other similar wrecks were known to lie in that vicinity. The nature and extent of the wreckage indicated a British naval vessel of the *Pandora*'s size and armament and the artefacts examined were all from the late eighteenth century, the period the *Pandora* was lost. A maker's name on bronze fittings matched that of a contractor known to have been supplying such fittings to the yard where the *Pandora* was built.

In my report I concluded that the wreck of the *Pandora* had very considerable historical significance. The mutiny on the *Bounty* and its remarkable train of events was one of the best-known sea stories of all time. Certainly it was the most romanticised. The mutiny had made the world keenly aware of the South Seas, and since then the episode had been the subject of numerous Hollywood films and books. No other Australian sea stories—even the grotesque experiences of the *Batavia* survivors—had aroused international interest comparable with that elicited by the romance and tragedy of the *Bounty* mutiny and its consequences.

Like the Dutch wrecks on the West Australian coast, the *Pandora* played a role in European (or perhaps global) rather than Australian history. The Dutch ships were lost while on a passage from Holland to the East Indies, and the *Pandora* was wrecked while heading home to Britain from Tahiti. In neither case was the vessel intending to pursue any intercourse with Australia. The broad historical significance of the *Bounty* episode was to be measured in terms of the European penetration of the South Pacific—what Alan Moorehead has termed 'the fatal impact'.[7] That penetration commenced with the arrival of James Cook in the *Endeavour*, was accelerated by Fletcher Christian and his band of mutineers, and seemed complete when the economic exploitation of the region reached its peak in the nineteenth century. And perhaps Britain's old rival, the French, will yet take the ultimate step.

In assessing the archaeological significance of the *Pandora* wreck, two aspects were considered: the hull and equipment belonging to the ship; and the cargo and crew's possessions.

The observed distribution of material exposed on the seabed indicated that the vessel was subjected to very little disturbance after settling on the bottom. Judging from the surface layer inspection of the wreck, and the site conditions—deep water, level sand, no rock and no turbulence—the *Pandora* might prove to be one of the best preserved shipwrecks in Australian waters. Certainly the conditions for preservation were more favourable to those on any other known pre-1800 wreck.

The *Pandora* wreck fits comfortably into Keith Muckelroy's criteria for a first-class condition site. The actual sinking circumstances of the *Pandora* were somewhat similar to those of the *Wasa*: in each case the ship filled with water, sank to the bottom virtually intact, and remained there relatively undisturbed. The *Wasa* had the additional advantage of a teredo-free mud bottom, which enabled the survival of the upper parts of the vessel.

Assuming for the moment that the *Pandora* hull is in an excellent state of preservation, what is its potential as a medium for the study of eighteenth century shipbuilding? Already during the eighteenth century the use of plans had become more widespread. The English Navy Board required construction drawings from 1716 on, and these became increasingly comprehensive as the century progressed. The plans used to build the *Pandora* survive in the Admiralty collection at the National Maritime Museum, where detailed specifications and fine contemporary models of similar ships can be found. As far as the archaeological remains of these vessels are concerned, the importance of the physical evidence lies in studying the finer details of ship construction, and in ascertaining whether all of the naval specifications were followed, or if they were modified in the course of the vessel's life. The well-documented *Pandora* provides a good opportunity for such research.

Britain has a bigger eighteenth century warship—HMS *Victory*. But the *Pandora*'s scale is not comparable with that of *Victory*. Earlier, bigger, and probably better-preserved warships lie on Britain's Goodwin Sands.[8] The USS *Constitution*, a frigate built in 1797 and restored in the United States, is also considerably larger. The raised hull of the *Pandora* would provide the means for detailed comparison with these more substantial warships and an avenue to study the smaller ships with which Britain ruled the seas. She represents the ordinary working warship of the Royal Navy—not an outstanding design, but a useful vessel built in numbers and used for all sorts of second-line tasks, differing only in size, strength and number of decks from the larger vessels which fought in the line of battle.

The condition of the wreck favours the survival of both the cargo (in this case the breadfruit plants and the many stores intended for the *Bounty* and her crew), and the hastily abandoned *Pandora* crew's possessions. Deeper levels within the wreck might yield some of the wooden containers of the breadfruit. Other aspects related to the particular nature of the *Pandora*'s mission could be explored. The Admiralty intended that the *Bounty* be recaptured and returned to England, so the *Pandora* carried a full complement of naval stores to refit the *Bounty*. These stores, together with the breadfruit plants, would make for a grossly overcrowded ship. *Pandora* also carried a *Bounty* anchor left by the mutineers at Tahiti, clothing and buttons left by the French explorer La Perouse at Vavua Island, an axe left by Captain Cook on Anamooka Island, and Polynesian artefacts traded with islanders. With careful excavation, study can go beyond the technological features of the vessel to social aspects of the crew's accommodation and their activities.

In view of these considerations, I recommended to the Department of Arts, Heritage and Environment that since the position of the *Pandora* site was known it should be excavated. The West Australian experience had shown that only archaeological excavation of important shipwrecks in isolated areas could save them from destruction by looters with explosives. It would not be an easy task. The wreck is 112 kilometres offshore and 645 kilometres from the closest city. Treacherous reefs, cyclones and stinging jellyfish are but some of the hazards of the environment.

The first excavation season was conducted during October and November 1983. The expedition was organised by the Queensland Museum and the Department of Arts, Heritage and Environment, but in terms of personnel it was very much a national expedition, with members drawn from Western Australia, Tasmania, and South Australia, as well as Queensland. Funding came principally from the Commonwealth Government.

As it was the first excavation season on what was known to be a difficult, extensive and isolated site the principal objectives were:

(a) to get to know the site and its environment, and to establish working procedures both above and below the water;
(b) to establish a survey control system that would be effective for the duration of all excavation seasons on the site; and
(c) to open several small exploratory trenches to help in assessing the condition of the hull and the quantity of artefactual material to be expected from future excavations;
(d) to raise from these trenches a limited quantity of diverse material to give conservators some appreciation of the types of conservation problems they would be facing subsequent to future excavations, and to give archaeologists an indication of the range and condition of artefacts on the site.

The 22 personnel leaving Cairns consisted of the project director (Ron Coleman), the archaeological director (myself), five other archaeologists, the dive master, the archaeological photographer, the field conservator, the diving doctor, the re-compression chamber operator, five general hands, and a crew of five. The expedition spent 27 days in the vicinity of the wreck site.

Flamingo Bay, a 23.5-metre ex-prawn-trawler, was the expedition work boat. This vessel has a two-man twin-lock re-compression chamber, a three-phase generator, two compressors, ample lifting facilities and good accommodation.

On arrival at the site, marker buoys were laid at the wreck's extremities, and the mother vessel was anchored fore-and-aft over the site, enabling it to stay overnight and avoid losses of diving time. The depth of water limited the daily diving time to one 18-minute dive at the beginning of the season. The problem of obtaining adequate archaeological direction on the seabed was resolved by forming a number of dive teams, each under the leadership of a qualified archaeologist.

Team leaders were individually briefed during the afternoons on the tasks intended for their team the following day. It was then the team leader's responsibility to draw up on an underwater slate a step-by-step detailed plan of the tasks to be

A photomosaic of the *Pandora* site, with detail showing cannon and chain pump fittings. (Photo: Pat Baker)

carried out during the morning and afternoon dives. The daily briefing, attended by all diving personnel immediately after the evening meal, was intended more to let team members know what other teams would be doing than to give detailed instructions on each diver's tasks. These evening meetings were also de-briefing sessions, and acted as a sounding board for alternative approaches to particular problems.

The photomosaic obtained during the 1979 expedition was used as a guide to establish a network of aluminium survey poles, spaced 10 metres apart, throughout the extent of the wreck site, so that provenance could be recorded. It was necessary that these poles be hammered into the seabed so firmly that they would endure for the duration of all excavation work that might take place in the future.

A thin wire rope was laid out and tensioned to produce a straight line in plan view. This base-line ran past the wreck, parallel with the supposed line of the keel. Then 3-metres-long poles were hammered 1–1.5 metres into the seabed at 10 metre intervals along the base-line. A closed iron sleeve was placed over the top of the pole and used as a pile driver. At the seabed level an open sleeve, mounted on a heavy levelling platform, guided the pole vertically into the seabed. For correct spacing and alignment of the poles on the baseline, several perpendicular open sleeves were bolted onto a horizontal bar at regular intervals and slipped over the poles being hammered into position.

Once the 50-metre base-line was established it was used to construct a second, parallel line through the site. Wire bridles forming right-angled triangles were used for pole placement. Eventually the area covered amounted to 50 metres by 20 metres, or 1,000 square metres. A datum mark was placed on each of the 18 control poles by placing a horizontal bar (surmounted by a level) against two poles at a time and transferring the height from one to the next pole. Then X, Y and Z co-ordinates could be used for artefact provenance.

Three small trenches were dug to discover the orientation of the buried hull and to expose some ship's fittings, stores and crew possessions, so that indications could be obtained of the volume, condition and range of artefacts to be expected on the site. Several small moveable grids were constructed for these trenches.

The first trench encompassed the exposed sternpost sheathing. The wood of the sternpost itself had been destroyed by marine organisms, but the sheathing retained the form of the post. The intention was to follow the sternpost sheathing downwards to where it joined the aftermost point of the keel. Then one end of the ship would be accurately located, and the vessel's heel or tilt on the seabed could be recorded.

The first important details to come to light were Roman numerals—'XIII', 'XII', and 'XI'—applied in lead to the side of the sternpost. These numerals indicate the ship's draft in feet, and are measured from the bottom of the keel. We knew then that we should be 11 feet (3.4 metres) from the bottom of the keel. But as we followed along the line of the sternpost to about the 9-foot mark, the sheathing lost its hitherto clear form, and the trench was abandoned when the sternpost-keel junction was not located at its predicted location.

We knew that the rudder and sternpost were badly damaged when the *Pandora* struck the reef; Captain Edwards referred to part of the sternpost being beaten away on the reef, and divers found rudder fittings on top of the reef in 1977 and 1983. So it is now clear that the keel itself will not be encountered until the trench is extended forward beyond the area of damage. However, examining the site we noticed the sheathing of the sternpost extended forward into the ship, suggesting that the upper sternpost remained intact with the hull after the sinking, and the sternpost tilt indicates that the ship settled on her starboard side.

The second trench was intended to reveal the forward extremity of the keel at the juncture with the stem. Again, the sheathing had retained its form as an intact skin, and the timbers of the stem (at the uppermost levels at least) had disappeared. A large quantity of ship's stores—in particular some large earthenware oil jars and a number of spare rudder fittings—slowed the excavation of this trench, and its full potential was denied because of insufficient time. However, well-preserved timbers appeared below the earthenware jars.

A third trench, 2 metres square, was dug to clear a partially exposed cannon for raising. This trench was not extended beyond the depth necessary to free the cannon without endangering surrounding material. It contained the heaviest concentration of artefacts, some 200 items being recovered. These consisted of ship's stores, personal effects and professional items related to the surgeon's practice.

Artefacts recovered from the sternpost trench included a large glass goblet presumably related to the officers' accommodation, three small orange-brown glazed bottles similar to ink bottles of a later period, a bone fishing lure possibly of Polynesian origin, bronze fastening bolts stamped with the broad arrow of the Admiralty and in one case the name 'Roe and Co.', and a number of small clear-glass panes each bearing the broad arrow mark.

The forward trench yielded two large earthenware oil jars, and the remains of several pulley blocks.

The surgeon on board the *Pandora* was George Hamilton, whose journal of the voyage has been published. The third trench revealed a number of items possibly associated with his practice. These include a brass tourniquet clamp, a brass case for surgical instruments containing a small stoppered glass vial, an ivory syringe, a marble mortar, the remains of pill boxes, a variety of stoneware jars, some of which retained their contents, and glass bottles and jars containing basic medicinal compounds such as oil of cloves, and other substances including calcium hydroxide and lead sulphide. Personal items from this trench include a fob watch with silver case and gold parts, bone or ivory buttons, a dinner plate bearing the incised initial 'W' or 'M', a lead pencil, a lice comb, and several brushes, one with a shoe-horn fitting.

This trench also yielded ship's stores, including large numbers of case bottles and cylindrical spirits bottles, wooden barrel staves, and a copper barrel-hoop. Fittings from the ship included cabinet knobs, a lock plate and key hole, and a red clay hearth brick. The ship's defences were also represented. Besides the iron 6-pounder cannon there were 112 lead balls for small-arms ammunition, one canister of shot, a trigger guard from a flintlock pistol, and a butt plate from a flintlock musket.

The second excavation season, in 1984, saw the commencement of systematic

A remote-piloted vehicle films divers using a water dredge to excavate the stern section of the *Pandora* wreck. (Photo: Pat Baker)

excavation through the 1,000-square-metre area gridded the previous season. A series of interlocking two-metre-square aluminium grids was laid out across the site as a guide for the excavators. The 1,000-square-metre area would hold 250 of these grids.

We began our digging at grid 1, behind the ship, and systematically moved forward, past the sternpost and inside the ship. We were able to start to quantify our progress. By the end of the season we had completed work on 65 grids—a quarter of the site. But the wreck site is complex. As we move towards the centre of the ship the deposit gets deeper, which means that each grid will take longer to excavate.

Towards the end of the 1984 season we came across intact hull planking and frames less than 1 metre below the seabed, in the area of the ship's breadroom. Admiralty draughts of the *Pandora*'s lines make it clear that forward of this section the

A cabin fire-place, chimney intact, is exposed on the *Pandora* wreck. (Photo: Brian Richards)

planking will curve more deeply into the sand, so it is apparent that a large proportion of the lower wooden hull has survived.

Acoustic equipment (an ORE 1–12 KHZ tunable profiler) was used to obtain a series of sub-bottom images of the hull remains lying hidden below the seabed.[9] The sub-bottom profiler utilises low-frequency sound to penetrate bottom sediments. This information suggests to the operators, from James Cook University, that the hull extends 5 metres into the seabed—a little more than the depth indicated by the archaeologists' physical examination of hull timbers at the stern.

Many people ask how long the *Pandora* excavation will take to complete. If sufficient improvements are made in mooring and lifting facilities, then it should be possible to use the 1984 season—when 25% of the site was opened up—as a guide. This would imply 3 more seasons to complete the detailed survey of the hull, and to raise the artefacts from within the hull. By that time it should be clear as to whether raising the hull itself would be useful or possible. The raising of the hull itself, however, would be a completely different operation from excavating the contents.

Among the finds during the 1984 season were an iron swivel gun, the barrel of a blunderbuss, and a variety of South Seas artefacts, including beads, cowrie ornaments and bone fishing lures. But the most spectacular individual find of the season was a cabin fireplace, which would have been used to heat an officer's cabin and dry his clothes. The iron fireplace was recovered intact, ornamented with an impressed façade and a copper vent.

This array of artefacts in an impressive state of preservation fully confirmed earlier predictions about the site. It is clear that excavation will shed light on the way of life of sailors and officers in the eighteenth century British Navy in a way not previously possible by archaeological means.

In some respects the *Pandora* excavation is quite unlike previous shipwreck excavations in Australian waters: the principal players in the *Bounty* mutiny story are so well known. The propaganda firstly of a self-interested British upper class and then of Hollywood film producers has selected and magnified particular character traits of these men so effectively that inevitably many of us see these larger than life images as characterising European and Polynesian society at that time of early contact. William Bligh and Edward Edwards are generally personified as the bullying, unfeeling commanders; Fletcher Christian, Peter Heyward and their fellows are restless spirits, agonising over the pros and cons of two disparate cultures; and the ordinary crewmen—are just unthinking rabble. The Tahitian women are beautiful, fiery and wanton—they are the noble savage. The peoples of several islands—in particular Pitcairn and Norfolk—look back to the mutineers and their wives as their progenitors. For the Pitcairners and Norfolk Islanders, the image of the *Bounty* mutineers is of particular consequence.

Excavation of the *Pandora* is unlikely to change our interpretation of the major events. But it will help to bring the story down to earth by showing more of how the men looked and lived, when working, eating, playing or sleeping on a British warship in Polynesia during the *Bounty* episode. The tools of Surgeon Hamilton can perhaps say as much about his role on board as does his diary. The tools of the ordinary non-writing crewman will, in the same manner, give him a voice for the first time.

2. Sites Protected by the Historic Shipwrecks Act, 1976

Among the Queensland sites protected under the Historic Shipwrecks Act are the material thrown overboard from HMS *Endeavour* on 12 June 1770 when she was stranded on a reef, the survey vessel *Mermaid* of 1829 (not yet discovered), the trader *Morning Star* of 1814 (again, not yet discovered), the steamer *Gothenburg* of 1875 (on which 102 passengers lost their lives), the iron barque *Scottish Prince* wrecked in 1887, the Royal Mail Ship *Quetta* of 1890 (on which 133 passengers lost their lives), the schooner *Foam* of 1893, the iron barque *Aarhus* of 1894, and the steamer *Yongala* of 1911 (on which all 120 occupants perished).

The records show that Queensland has a large number of early colonial period shipwreck sites, thanks to the Great Barrier Reef and Torres Strait. However, it is several of the later sites, in particular the *Foam* and the *Yongala*, that have received attention from the Museum.

Ceramic amulets—trade goods being carried on the Queensland labour trade vessel *Foam* when she sank in 1893. (Photo: Jon Carpenter)

The schooner *Foam* was heading for the Solomon Islands to return labourers who had completed their indenture period when she struck Myrmidon Reef north of Townsville. The crew and 84 Kanakas, returning from the Queensland sugar industry, were saved. The wreck, since discovery in 1982, has been found to contain artefacts associated with the Queensland labour trade, commonly referred to as blackbirding. Artefacts raised from the site include dozens of porcelain armlets—European-made trade goods intended to imitate the traditional ones carved from giant clam shell by South Sea Islanders.[10] This sort of trade item has not previously been seen in Australia during modern times.

The SS *Yongala*, one of the Adelaide Steamship Company's most popular and luxurious coastal steamers, was lost in a cyclone while *en route* from Melbourne to Cairns with 48 passengers, 72 crew and 686 tons of general cargo. The wreck was declared of historic significance under the Historic Shipwrecks Act in June 1981, and Queensland Museum staff examined the site in August of that year. A special zone was declared around the *Yongala*, prohibiting trawling, fishing, spearfishing or other collecting of marine life, as well as anchoring directly over the wreck. The hull lies in deep water and is reported to be virtually intact.[11] If it can be adequately protected, the site will offer a very valuable resource for future archaeologists examining the way of life on a coastal steamer of the early twentieth century. A site such as the *Yongala*, where iron bulkheads and components still provide complete separation between different parts of the ship, provides a good opportunity for archaeologists to seek answers to specific research questions without excavating the entire area of the wreck site on an open plan excavation strategy.

It is clear that unless the State Government recruits conservators, planners and researchers for the Maritime Archaeology Department the continued *Pandora* excavations will soon bring about a crisis situation. The *Batavia* excavations in Western Australia were serviced by a team of some 15 conservators, and the volume of artefacts resulting from a total excavation of the *Pandora* could be several times greater than that from the *Batavia*.

Chapter 8 REFERENCES

1. Agnew, 1983, p. 110.
2. Henderson, 1980, p. 239.
3. Edwards, in Thomson, 1915.
4. Ashdown, 1972.
5. Chapman, 1775, p. xxxi.
6. Greenhill and Lyon, 1979, pers. comm. of material from ADM/A/2931, NMM.
7. Moorehead, 1966.
8. Personal communication from David Lyon.
9. Johnson and Hooper, 1984.
10. Personal communication from Ron Coleman.
11. Coleman, unpublished report.

9 Maritime Archaeology in Tasmania, Victoria, South Australia, New South Wales and the Australian Territories

1. A Rum Trader in Tasmania

Tasmania, like Queensland, was spurred into an involvement with maritime archaeology as a result of the discovery of a particular wreck site—in this case the *Sydney Cove*.

The wreck was found in January 1977 by a group of divers making a film about wrecks around the islands in Bass Strait. Matthew Flinders had marked the position of the wreck on a chart when he surveyed the area in 1798, but the divers found the site by chance. They raised the rudder (complete with its bronze pintles), wood samples, pieces of cane, lead fragments, broken bottles, the top of a barrel and two pulley sheaves. The rudder went to the Queen Victoria Museum and Art Gallery, via the Receiver of Wrecks.[1] In the absence of specific historic shipwrecks legislation to protect the site (recognised immediately as being one of considerable importance), the area in which the *Sydney Cove* wreck lies was declared in March 1977 to be a State Reserve under the National Parks and Wildlife Act of 1970.

The interest generated by the finding of the *Sydney Cove* led to a number of developments in Tasmania. A divers' underwater research group was formed, and as a result of the organisational efforts of Ken Atherton, one of the finders, the research group became the Maritime Archaeological Association of Tasmania in 1978. The Association has since been engaged in drawing together a sites' register (on computer at the Queen Victoria Museum), has carried out searches for sites such as the convict-built brig *Apollo* lost off the northern end of Maria Island in 1827, and has surveyed underwater areas adjacent to sites of early convict settlements, such as that at Mason's Cove, Port Arthur.[2]

Association members have played an important role in the several expeditions to the *Sydney Cove* site, providing manpower and equipment, and doing much of the organisation. They also lobbied for the initiation of a programme of maritime archaeology at an appropriate institution.

Two institutions have become involved in maritime archaeology in Tasmania: the Tasmanian National Parks and Wildlife Service, based in Hobart; and the Queen Victoria Museum and Art Gallery in Launceston.

The National Parks and Wildlife Service is a relatively large State Government organisation (with a staff of some 90–100) whose function includes the management of prehistoric and historic sites in Tasmania. It has been involved in prehistory and historical archaeology for some time, and became interested in maritime archaeology when the National Parks and Wildlife Act was used to protect the *Sydney Cove* wreck in 1977. A grant of $10,000 was made available in 1978 by the State Government through the National Parks Service to fund an expedition to the *Sydney Cove* wreck. In 1979 an archaeologist was appointed with a view to some involvement in maritime archaeology. The National Parks Service organised another expedition to the site of the *Sydney Cove* in 1980, when relics were thought to be in jeopardy. It was then able to supply some of the expedition equipment, including the National Parks Service landing craft.

The Commonwealth Historic Shipwrecks Act was proclaimed in Tasmania in 1982, and the *Sydney Cove* wreck was the first Tasmanian site to be protected. Responsibility for the implementation of the Act was delegated to the National Parks Service. A maritime archaeologist (Paul Clark) was appointed temporarily to the National Parks Service in 1984. He led a brief survey expedition to the *Sydney Cove* wreck soon afterwards, and has since been able to commence a programme of official inspection of known sites around the Tasmanian coast.

The Queen Victoria Museum and Art Gallery, with a staff of around 40, is responsible to, and principally funded by the Launceston City Council. The Museum became involved in maritime archaeology when it accepted the rudder raised by the *Sydney Cove* wreck finders in 1977, and since that time all artefacts removed from the site have gone there. A conservator was appointed in 1979 to deal with the *Sydney Cove* collection. The laboratory has a freeze dryer for the treatment of organic materials, and facilities for the treatment of iron. The conservator until late 1985, Shirley Strachan, also performed for a time the role of maritime archaeologist within the Museum, and as a part of that role drew up a research design for future work on the *Sydney Cove* wreck.

Preservation Island lies among the Furneaux Group at the eastern end of Bass Strait. The Island is leased to a fisherman-farmer who runs a few head of cattle and does some muttonbirding. The Island is accessible by light aircraft and boat.

The *Sydney Cove* wreck is located between Preservation Island and Rum Island, in a bay sheltered from the north and the west. Ocean swells and tides occasionally affect the site. The wreckage is buried in sand on a gently sloping bottom in 3 to 4 metres of water. Part of the site is covered with sea grass, and this stabilises the sand in the area.

Archaeologists visited the site in 1978, 1980, 1984 and 1985. I joined a week-long 12-man expedition in 1978, to give archaeological direction. The aim of this feasibility study was to ascertain what proportion of the ship had survived, by exposing the extremities of the upper surface of the keelson (the piece which locks the ribs of a ship on to the keel) and by remote sensing of buried material.[3]

Hookah-breathing divers operated a water dredge to clear sand from above the keelson. As the keelson was followed from the bow towards the midships area of the

A landing craft, with its drop bow, was used as the on-site work boat during a feasibility study of the *Sydney Cove* wreck. (Photo: Graeme Henderson)

wreck, the sand deposit deepened, and the number of loose artefacts increased, so the dredging was ceased at the main-mast step. Basic structural features forward of the main-mast were trilaterated.

Beyond the confines of the hull, a sea-going magnetometer was used to search for iron concretions. To prevent metallic interference from personal equipment, a diver swam along on the water surface and dangled the magnetometer 'fish' by its lead just above the seabed for each recording. The recordings flashed on to a digital screen on the above-water console at 2-second intervals and were noted on paper by the operator. Areas of anomalies (variations in the measured magnetic field) were then marked on an overall site plan for future investigation.

A magnetometer survey was also conducted on an adjacent area of Preservation Island, at a place commonly referred to as 'the fireplace'. Locals believe that the *Sydney Cove* crew camped here between the sheltering rocks, and built a house from parts of the ship. Again, anomalies were marked for future investigation.

Another week-long expedition was arranged in 1980 with the intention of raising the remaining cannon and anchor (judged by the National Parks Service to be under threat from souvenir-hunting divers), and completing the earlier aim of establishing the aftermost extent of the keelson. To minimise disturbance of the area aft of the main-mast step, a trench was dug at right angles to the line of the keelson, in the predicted location of the sternpost. Subsequent trenches dug further forward on the wreck located a sternpost gudgeon and a second mast step, but the aftermost extent of the keelson remained obscured by planking, dunnage (wood packed around cargo to prevent movement and damage) and other material which had collected in the hold as the ship broke up in 1797.

All that the currently available documentary evidence tells about the *Sydney Cove* itself is that it was a two-decked ship. Nothing is said of the tonnage, dimensions, or place and date of building. But the archaeological evidence gleaned from the expeditions to the wreck enable some preliminary observations to be made about its structure.

It was a three-masted vessel. Only the main and mizzenmast steps remain but their positions on the keel suggests that the vessel also had a foremast. The extremities of the keelson have not been located, but the observed positions of the mast steps suggest that the *Sydney Cove* had a keel length of around 25 to 27 metres, and therefore a deck length of some 28 to 30 metres. The suction pump was uncovered on the site and measured some 6.5 metres, so the depth of hold in the *Sydney Cove* was around 6 metres.[4] None of the archaeological information so far obtained gives any indication of the *Sydney Cove*'s beam. However, comparing the vessel with other contemporary two-deckers of similar length indicates a vessel of 250 to 350 tons. Tonnage measurement at that time was less meaningful than it is today because various methods of calculation were used at random, and depth of hold was not taken into proper account.

It is illuminating to compare and contrast the hull of the *Sydney Cove* with the hulls of two later shipwrecks which have been carefully investigated in Australian waters: the 366-ton two-decked American China-trader *Rapid*, built in 1807; and the 167-ton French 1830s-built one-decked slaver *Don Francisco* (later renamed *James Matthews*).[5]

The two-decked *Sydney Cove* was lightly built—in many respects no more heavily built than the 26-metre one-decked *Don Francisco*, and much lighter than the 32-metre *Rapid*. The forward frames were single (near the keelson at least), and widely spread. At the keelson the sided width of the floor timbers of the *Sydney Cove* was 21 centimetres, compared with 16 centimetres on the *Don Francisco* and 29 centimetres on the *Rapid*. But the *Don Francisco* and the *Rapid* were both double framed, which meant that effectively each frame was 32 centimetres sided on the *Don Francisco*, and 58 centimetres sided on the *Rapid*, compared with 21 centimetres on the *Sydney Cove*. The space between frames on the *Sydney Cove* was 39 centimetres, compared with 16 centimetres and less on the *Don Francisco*, and 2 centimetres or less on the *Rapid*. So the loose framing of the *Sydney Cove* contrasts markedly with that of the *Rapid*.

Given that the *Sydney Cove*'s ribs were narrow and widely spaced it might be expected that the floors—the rib sections passing under the keelson—would all be continuous for the flatter sections of the hull, to maintain transverse strength. But this was not so. Some floor timbers were joined by a scarf joint extending over the

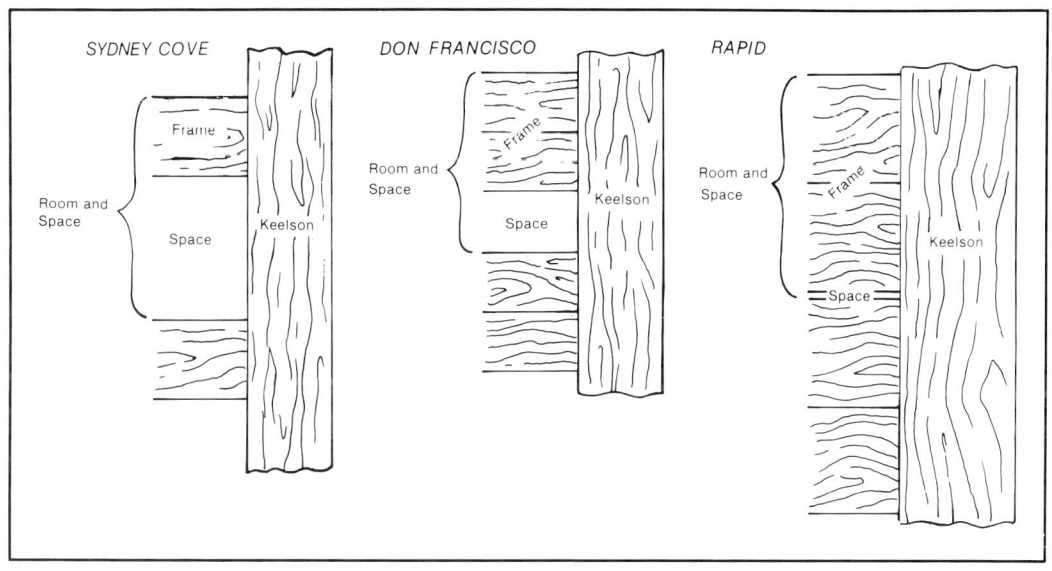

The *Sydney Cove* is lightly built when compared with Atlantic-built vessels of the early nineteenth century.

The frames of the heavily built *Rapid* are packed tightly together. (Photo: Pat Baker)

keel, with a gap in the lower half of the joint to leave room for the keel. The frames of the *Rapid* and the *Don Francisco* regularly alternated between continuous floor, and rib butting together beneath the keelson.

The *Sydney Cove* had no ceiling (inside lining) planking. The thickness of the bottom (outside) planking was 7 centimetres. Both the *Don Francisco* and the *Rapid* had much thicker hulls. The *Don Francisco*'s bottom planking was 9 centimetres and her ceiling planking 7 centimetres, making a total thickness of 16 centimetres. The *Rapid*'s bottom planking was 10 centimetres thick and her ceiling planking 5 centimetres thick, making a total thickness of 15 centimetres. In later times, the lack of ceiling planking would limit the types of cargo a vessel was allowed to carry.

The *Sydney Cove*'s keelson was 24 centimetres wide, compared with a 23-centimetre *Don Francisco* keelson, and a 34-centimetre *Rapid* keelson. For much of its length the *Sydney Cove*'s keelson was surmounted by a rider keelson—a longitudinal beam of the same width as the keelson—and this piece would have compensated to some extent for the narrow keelson. All of the abovementioned features considered together, however, indicate that the *Sydney Cove* was much lighter in build for its size than either of the other ships mentioned.

The *Rapid*, built in Boston for the China trade, was a strong, full-built ship designed to carry cargo relatively quickly half way around the globe, through punishing weather in southern latitudes. The *Don Francisco*, built in France to carry slaves during an era of prohibition, needed to be strong, sharp built and relatively shallow in draft, for speed to elude capture in the Atlantic, and the ability to hide in mangrove shallows. What sort of vessel was the *Sydney Cove*? Unfortunately we do not yet know with any certainty where she was built, or for what purpose. If, as seems likely, the *Sydney Cove* was built in India or thereabouts, then she would probably have been intended for the country trade, ranging about the Indian sub-continent and adjacent waters, but not for long ocean voyages.

It comes as no surprise then that a country trade vessel would encounter difficulties on such a long uninterrupted voyage as that to New South Wales, taking the ship through the savage gales of the southerly latitudes. The trade with New South Wales came as a sudden new development which may well have caught the merchants involved unprepared in terms of suitable available ships. The seven ships which came from India to the Colony prior to the *Sydney Cove* ranged from the tiny rum-trade brig *Arthur*, of 95 tons, to the 800-ton stock carrier *Endeavour*, but the majority were small vessels.[6] The first vessel despatched from India to New South Wales (other than English convict transports) was the *Neptune*, procured after delays by Captain William Bampton in fulfilment of his contract of early 1793 to supply the Colony with much-needed cattle and grain. But the *Neptune* was lost soon after her departure from Bombay. No other vessel was available in India until May 1794, when the ship *Endeavour* arrived, and was purchased by Captain Bampton. Extensive delays followed, partly because of the necessity of having to dock and refit the *Endeavour* (said to be 70 years old at the time) for the voyage. At last the vessel sailed for Sydney in March 1795. The *Endeavour* left Sydney the same year for New Zealand. On the 4th October she encountered a heavy gale which opened up her decayed seams to the extent that the pumps had to be kept going continuously for 2 days. She was in such bad condition that she was scrapped on arrival there, and became New Zealand's first recorded shipwreck.[7]

Aspects of the *Sydney Cove*'s manning and equipment also indicate a vessel poorly prepared for this long speculative voyage. The Indian sailors, unused to the cold climate, collapsed at the pumps, and three of them died there. The *Sydney Cove* was furnished with common pumps, operated by a lift action on deck and representing back-breaking labour for the crew. There are no signs on the wreck that the vessel was equipped with the more efficient chain pumps, which produce a continuous flow. The two small cannon (2 or 3 pounders) found on the wreck would have been inadequate defence against serious attack at sea in the 1790s, but it remains possible that more guns await exposure on the site.

The raising of objects from the *Sydney Cove* wreck site during the 1978 and 1980 expeditions was restricted to several relics thought by the National Parks Service and the Museum to be otherwise liable to theft by souvenir hunters, and to some sample items to assess the condition and significance of the wreck. Three anchors (two with wood stocks and one with an iron stock) and two cannon were exposed on the site. The guns and two of the anchors were lifted and taken back to Launceston for storage and conservation.

Other items raised from the site included spirit or wine bottles (one still full when taken ashore and bearing its scarlet wax seal with the initials 'C.C.' of the consignors Campbell, Clarke and Co.), pieces of wooden barrels, a small sand glass (for timing the log line) complete with its sealing cork and fine grained sand, cannonballs, leather shoes, fragments of Chinese porcelain bowls, plates and jars, bullock and pig bones, an ebony-handled knife, an earthenware jar and bowls, bamboo fragments, and some small wooden logs which would have been used as dunnage.

What then is the historical significance of the wreck? The *Sydney Cove* was the first merchant ship to be wrecked in Australian waters after the establishment of the colony of New South Wales. The cargo of rum carried on board the *Sydney Cove* constitutes the earliest form of extensive commercial activity carried out in the colony. The rum imported from India and elsewhere was the centre of far-reaching social, economic and political changes that were occurring in New South Wales at the time. The wreck itself is representative of the very beginnings of trade between Australia and the outside world, and it was the development of trade which enabled Australia to move from its limited status as a convict outpost to become a colony of settlement, and eventually a nation.

The various sea journeys arising from the wreck of the *Sydney Cove* are also of considerable significance. These led to the recognition of Tasmania as being an island and to the identification of important navigable channels through the eastern end of Bass Strait. The knowledge of the presence of fur seals near habitable islands caused the first European settlement to be made in Bass Strait shortly after for the purpose of obtaining skins for export in accordance with an official direction to generate income within the new Colony.[8]

Thus the attention drawn to the area by the wreck of the *Sydney Cove* contri-

A sand glass, complete with cork, was found on the *Sydney Cove* wreck site.
(Photo: Queen Victoria Museum)

Ceramic plate, bowl and jar from the *Sydney Cove* wreck.
(Photo: Queen Victoria Museum)

buted to the establishment of Australia's first export industry, and to the foundation of settlement in Tasmania.

In assessing the archaeological significance of the wreck it is necessary to consider the hull and equipment belonging to the ship, the cargo and the crew's possessions, and the material left on the islands by the survivors who camped there.

The feasibility surveys established that a substantial portion of the hull has survived. The keel and keelson remain locked together for much of their length, holding in place the floors and futtocks, to which large areas of planking are still attached. Within Australian waters the *Sydney Cove*'s timbers and those of HMS *Pandora* are the only substantial eighteenth century ship structure known to have survived (other than eighteenth-century-built ships lost during the nineteenth century). Overseas, work has recently been carried out on a number of late eighteenth century shipwrecks, but few of these have involved substantial structure. During the second half of the eighteenth century, naval architecture was rapidly becoming a well-disciplined and documented subject in Europe, but the same does not necessarily apply to India, where it is likely the *Sydney Cove* was built, and where elements of traditional Indian shipbuilding may well have been incorporated in the ship. If it is definitely established that the vessel was built in India then the hull scantlings could be listed as a model, to look at features characteristic of Indian shipbuilding. The *Sydney Cove* has the advantage of being reasonably well preserved, but as a model there is the disadvantage that the India trade vessels seem to have been such a motley and varied group. A number of India-built ships were lost in Australian waters during the early colonial period, including the 317-ton *Valetta*, built in 1821; the 64-ton *Thistle* built at Bengal in 1825; the 444-ton *Cumberland*, built at Surat; the 83-ton cutter *Mermaid* built at Calcutta in 1816; the 350-ton ship *Mersey* built at Chittagong and wrecked in 1805; and the 207-ton *Emily Taylor*, built at Bombay in 1792. Investigation of these sites would provide comparisons with the *Sydney Cove*.

Then of course the *Sydney Cove* hull offers potential for investigation of aspects of the beginnings of India's trade with Australia. The *Sydney Cove* may be an example of a lag in adaptation by India's country trading commercial interests to new technological requirements of the longer Australian rum-trade route. Perhaps archival study of the later vessels employed in the trade would show an improvement in the quality of ships to have been prompted by the loss of the *Sydney Cove* (and part at least of its cargo and crew) and other worn-out vessels such as the *Endeavour*.

A number of the items of ship's equipment surviving in the *Sydney Cove* wreck, such as the pump, anchors and guns, have the potential for giving archaeological information. These delicate but bulky items demand considerable forethought if they are to survive excavation and conservation procedures.

The cargo items and crew's possessions lying on the *Sydney Cove* will, when fully excavated, constitute the only large reference collection of 'moveable' eighteenth century archaeological material directly relating to colonial Australia. And the *Sydney Cove* material is of course all accurately dated by the casualty. It will be useful to compare and contrast the origins and general nature of this collection with the collections excavated from around Sydney. Other than building foundations, very little archaeological material has survived from land sites of the foundation years. A preliminary article about excavation of the First Government House, Sydney, which is described as 'the most exciting archaeological find of European civilisation in this country', gives no mention of movable possessions found among the foundations.[9] Nor is it likely that substantial sources of such material will be found on other Australian sites in the future, so the *Sydney Cove* represents the one opportunity for a large collection to be made. Despite the fact that some cargo was thrown overboard from the *Sydney Cove*, and a great deal is known to have been salvaged by the crew, the surveys have shown that substantial quantities of ceramics, glassware and organic material have remained on the site. The sand has protected these items, and the exploratory trench dug across the stern in 1980 showed that more intact items are located under these deeper deposits.

On Preservation Island, isolated shards were seen lying at the mouths of muttonbird burrows over a large area, and the magnetometer survey indicated that more material lies buried in the area where local tradition has it that the survivors built their house. The island is the site of the first European occupation of Tasmania and as such has significance for Tasmania.

Items recovered from the *Sydney Cove* will be preserved and displayed in a maritime museum to be established in a restored warehouse building in Launceston with the aid of sponsorship from the LSB Statewide Bank, as part of the Bank's Sesquicentenary celebrations.

2. A Shipwreck Sites Management Programme in Victoria

The concept of maritime archaeology is a recent one in Victoria, where concerned divers first formed an Association in 1978. The event was timed to coincide with a project to raise an anchor from the wreck of the 1,693-ton iron ship *Loch Ard*, which sank with great loss of life on the south coast in 1878, one hundred years earlier.

At the time of the Association's formation no institution in Victoria showed any interest in maritime archaeology. This was a problem for the Association because any such amateur group needs support and guidance from professional archaeologists to achieve its full potential. Nor was there any legislation to protect maritime archaeological sites off the Victorian coastline. Association members had to contend with the spectre of the elusive Mahogany Ship at Warrnambool. Who could get excited about mid- to late-nineteenth-century sites when an ancient Spanish, Portuguese, Dutch or even perhaps Chinese vessel lay on the verge of discovery somewhere between Port Fairy and Warrnambool? And, for those interested in nineteenth century sailing ships, the comprehensive restoration project on the Melbourne water-front of the lovely iron barque *Polly Woodside* (a typical ship from the Golden Age of sail) represented more tangible evidence to most people than the weed-covered remains to be found on shipwreck sites.

Against these odds, the Maritime Archaeology Association of Victoria developed a useful role. Their sites' register soon listed some 500 shipwrecks, including 135 in

Museum visitors Kandy and Rae Hodgson by the helm of the *Polly Woodside*. (Photo: Graeme Henderson)

Port Phillip Bay.[10] Members consulted with the general diving fraternity to get accurate information on the locations of known sites, and familiarised themselves with some of the procedures for working on the seabed by conducting surveys on several shipwreck sites. One of the sites surveyed was the immigrant ship *New Zealander*, lost at Portland Bay in 1853. Association members carried out a comprehensive archival study to compile an historical background to the vessel.[11]

Association members (notably Terry Arnott) also played an important role in lobbying governments and institutions about the need for protective legislation. The Victorian Historic Shipwrecks Act was proclaimed in March 1982 after several years of planning by the Victoria Archaeological Survey (VAS) to ensure that it related effectively with the Commonwealth's Historic Shipwrecks Act, also proclaimed in Victoria in March 1982. The Victorian Act applies to Victoria's internal waters, such as Port Phillip Bay and Westernport Bay, while the Commonwealth's Act protects sites in Victorian coastal waters and adjacent areas not within the State's limits. Both Acts contain provision for the appointment of inspectors with powers of search and arrest. The Commonwealth and State Police enforce the Commonwealth Act, and the State Act is enforced by State Police, Fisheries and Wildlife Division (currently some 80 agents), National Parks Service and Ports and Harbours Division. The two Acts also make provision for declaration of Protected Zones around particular shipwreck sites if they are endangered.

Under the Acts the VAS is delegated the responsibility for their administration, for keeping a register of historic shipwrecks, and for carrying out appropriate fieldwork. The official repository for material from sites protected under the State Act is the Museum of Victoria.

The VAS, formed in 1972, is an agency of the Ministry for Planning and Environment. The main thrust of the VAS has been management of the Aboriginal cultural heritage—it is the official government agency responsible for the administration of Aboriginal sites in Victoria. This agency's recent development has been strongly influenced by the sorts of approaches explored in Schiffer and Gumerman's *Conservation Archaeology: A Guide for Cultural Resource Management Studies*.[12] Sections of the staff of some 30 persons deal with systematic survey, environmental assessments, the geology of sites, laboratory work, warden-inspection functions, and education.

In 1982 the VAS began to broaden the range of its responsibilities to cover historical archaeological sites. The Maritime Archaeological Unit (MAU) was started in September of that year as a result of preparatory work by current Director Mike McIntyre. The MAU now has three staff, the leader being State Maritime Archaeologist Mark Staniforth. Since the MAU commenced operations most of its work has been devoted to setting up an effective administration, but some fieldwork has also been possible. High priority has been given to the establishment of a computerised sites' register and inspection of sites for declaration under the Acts. Over 30 sites have now been inspected.[13]

Victorian sites protected under the Commonwealth legislation include the 64-ton schooner *Thistle*, which brought the first pastoral settlers to Victoria 150 years ago and was wrecked in 1836; the 255-ton barque *Children*, owned by James Henty and Co., lost in 1839 at Childers Cove to the east of Portland Bay; the steamship *Clonmel*; the 786-ton steamship *Monumental City*, lost near Gabo Island in 1853; the 2,284-ton Black Ball Line clipper *Schomberg* (one of the largest wooden sailing vessels built to that date), lost to the west of Cape Otway when 81 days out from England; the regular British trader *Loch Ard*; and the 405-ton barquentine *La Bella*, lost in 1905 near Warrnambool while carrying timber from New Zealand. The 251-ton brig *William Salthouse*, lost in Port Phillip Bay in 1841; the 67-ton New South Wales-built schooner *Clarence*, wrecked near the West Channel of Port Phillip Bay in 1850; the 160-ton brig *Mountain Maid*, a regular regional trader which sank after being run down by the steamship *Queen* in Port Phillip Bay in 1856; and the *City of Launceston* (a 368-ton screw-propelled intercolonial passenger-carrying steamship), lost in Port Phillip Bay in 1865 as a result of a collision with the steamship *Penola*, are protected under the Victorian legislation. The *Clarence, City of Launceston* and the *William Salthouse* sites are Protected Zones.

A number of sites have been examined with a view to further ascertaining their archaeological research potential. One of these is the 67-ton *Clarence*. Another site in the same area but of a different scale is that of the 368-ton steamer *City of Launceston*. Both sites present problems for survey or excavation because of the tide and resultant sand and weed movement.

A survey and test excavation was carried out on the *William Salthouse* wreck in March and April 1983, after it was found that divers had been smashing the cargo of barrels remaining on the site. During the excavation two trenches showed that some 3 metres of the hull had survived above the keel.[14] Under the confused upper layers of the site, intact barrels of provisions were seen remaining in their stacked positions. The brands and other marks on the barrel lids are providing useful information about the contents of the barrels, the date and place of packing, and the names of the merchants involved. These barrels are the only known large collection of nineteenth century provisions barrels in Australia, and as such present a new type of information resource. Because of the looting problem the *William Salthouse* had to be made a protected zone, but interested divers can obtain a permit from the VAS to look at the site.

It is clear that the direction taken by the MAU will follow reasonably closely that of the rest of the VAS—along the path of site management.[15] The VAS is an example of separated roles. It has responsibility for sites, but the Museum of Victoria has responsibility for any material raised, and the Polly Woodside Museum might accept material on loan from the Museum of Victoria for display. This basic separation of roles places a heavy reliance on goodwill between institutions if it is to succeed. The small staff at the MAU are establishing a reasonably large set of excavation equipment, including a 7-metre Sharkcat, hookah unit, high pressure compressor, water pump, metal detector, lifting bags, cargo nets and Nikonos cameras. But they do not have access to an expansive storage area, or a specialised conservation laboratory. Some items requiring conservation could be sent to other States where facilities exist (for example, leather to Tasmania), and consultant maritime archaeologists might be a way of organising the excavation of sites under dire threat of destruction. A $20,000 Heritage Commission-funded contract for a pre-disturbance survey of the *Clarence* wreck became available in July 1985, and a further $8,000 is scheduled for excavation work on the site. A number of Australian-built vessels are known to have been wrecked in the Bay, and the intention is to use this resource to build upon the body of knowledge pertaining to Australian shipbuilding. Another contract, funded by the Department of Arts, Heritage and Environment, involves a survey of wrecks on the west coast of Victoria.

3. Clipper Ships and Muddy Waters in South Australia

Maritime Archaeology in South Australia has been characterised by a long-term amateur interest and some support by the State Government.

The Society for Underwater Historical Research was formed in 1974, and initially had close links with the South Australian Museum. One of the first projects undertaken by the Society (now called the Maritime Archaeology Association of South Australia) was a systematic examination of the seabed at the site of the old Holdfast

Bay jetty at Glenelg, where passengers alighted from the first immigrant vessels and steam liners. All the different items which might be accidentally dropped from pocket or purse, or from shops on the jetty—coins, jewellery, pistols, watches etc.—have been found in the grid squares. The project was an opportunity for members to learn some of the basic techniques of working on the seabed. It also provided an insight to one of the character traits of landsmen: they have never learnt to adequately secure valuable possessions to their person.

A more ambitious project, supported by grants from the Australian Heritage Commission was centred on the old wharf facility belonging to the town of Morgan, on the Murray River. An airlift was used between grid lines in an area of no visibility, to assemble a collection of artefacts relating to the late-nineteenth-century paddle steamer trade. Worked glass tumbler bases and bottle bases found nearby were identified as tools made by the Aborigines.[16] Divers combatted the cold with hose-delivered hot water pumped into their wet-suits, and used equipment for communications in the murky water.

Surveys have been carried out on the shipwreck sites of the 225-ton snow *Tigress*, wrecked in the Gulf St Vincent in 1848 at the end of a voyage from Scotland; the 518-ton barque *Grecian*, wrecked at Port Adelaide's outer harbour at the end of a voyage from England in 1850; the 1,227-ton iron ship *Star of Greece*, wrecked at Port Willunga in 1888 with the loss of 17 of her crew; and the *Loch Vennachar*, a 1,552-ton iron ship wrecked at Kangaroo Island in 1905 with the loss of her entire crew. An anchor raised by the Society was treated by the Australian Mineral Development Laboratories, with funding from the Department of Environment and Planning who have provided both for the establishment of conservation facilities and for annual treatment work on large artefacts.

The first relevant protective legislation passed in South Australia was the Aboriginal and Historic Relics Preservation Act of 1965. The more comprehensive Commonwealth Historic Shipwrecks Act was proclaimed in 1980, and a South Australian Historic Shipwrecks Act was proclaimed in 1981. These last two pieces of legislation have identical protection and controlling provisions for Commonwealth and State waters.

Since 1977 the Heritage Conservation Branch of the Environment and Planning Department has steadily increased its commitment to the field of maritime archaeology. In 1980 the Branch sent one of its officers to Western Australia to do the Diploma in Maritime Archaeology, and that officer, Bill Jeffery, has since been the solitary officer responsible for maritime archaeology in South Australia.

In 1985 a substantial site inspection vessel was procured but given the lack of support staff and display floorspace (South Australia has only just acquired a State Maritime Museum), substantial fieldwork activities have not been entertained.
Rather, energies have been devoted to compiling a sites' register, and to protecting significant sites under the legislation.

In June 1985 South Australia's Historic Shipwrecks Act, 1981, was successfully tested. Five fishermen were charged in a Magistrates' Court with having been inside a Protected Zone without a permit. All were convicted and fined.

Some 340 ships are known to have been wrecked off the South Australian coast. The earliest ship known to have been wrecked was the barque *South Australian*, lost in Encounter Bay in 1837, and the earliest shipwreck located to date is the *Solway*, wrecked close to the *South Australian* a few days later. The *Solway* had brought German immigrants to South Australia from Hamburg. Only 5% of the South Australian wrecks have been accurately located, identified and assessed, so the register of sites is a primary goal.[17]

The first shipwrecks protected under the Commonwealth legislation were the *Loch Vennachar*; the three-masted 2,284-ton iron barque *Montebello*, lost in 1906; and the 3,596-ton turret-deck, screw steamer *Clan Ranald*, lost with 46 of her crew in 1909. The first declarations under the State Act protected the *Tigress*; the *Grecian*; the 762-ton Liverpool immigrant ship *Nashwauk*, lost in 1855; the 199-ton four-masted steel barque *Norma*, lost in 1907; and the 455-ton iron barque *Santiago*, built in 1856, abandoned in the north arm of the Port Adelaide River in 1945, and still basically intact.

Another 10 shipwrecks have been protected under the Commonwealth and State Acts since the first declarations in 1981. One of these was the remains of the 338-ton composite-built barque *Zanoni*, sunk in the Gulf St Vincent during a sudden squall in February 1867. The vessel was not found until 1982, and it is still relatively intact on a sandy bottom in 20 metres of water. A Protected Zone has been placed around the site to protect the fragile timber remains.

The Murray River has been an important area for transporting produce to and from the Australian interior. The first recorded European vessel to enter the Murray River through its mouth was the cutter *Water Witch*, which was wrecked at Blanchetown, hundreds of kilometres up the Murray, in 1842. The *Water Witch* was built by John Gray at Hobart in 1835 and came to South Australia as the Colony's first marine survey vessel in 1839. Under the command of Marine Surveyor Captain Pullen, she was used to chart the mouth of the Murray River and various bays in Spencer and St Vincent Gulfs. She was also used in assisting the explorer Edward Eyre when he crossed Australia from east to west. When the *Water Witch* sank in the Murray River in 1842, the Colony was in financial difficulty, and no funds were committed to raising a vessel said to be in poor repair.

The remains of the *Water Witch* have been protected under the State legislation, and an excavation took place as a 150th Jubilee project. Although a major part of the vessel has disappeared (the Murray River used to dry in various places and it would have been easy to salvage parts from her if she became exposed) enough still exists to document aspects of how the vessel was built. The *Water Witch* remains are an important link with some of the earliest shipbuilding carried out in Australia. It would be interesting to compare aspects of the *Water Witch*'s construction with other Hobart-built vessels of the same period. The remains of the brigantine *Miranda*, built in 1846, wrecked in 1852, and now lying in the intertidal zone in Miranda Bay at Wilsons Promontary, provides one such opportunity for comparison.

The excavation of the *Water Witch* was conducted in nil visibility using an

The hull of the *Miranda* lies buried on the beach at Wilson's Promontory. (Photo: Victoria Archaeological Survey)

aluminium grid from which were recorded the three-dimensional co-ordinates. This was carried out using tape measures and a torch to illuminate distances for the diver. Underwater communications were used to convey the measurements to a recorder on the river bank, and elements of the vessel's construction have been recovered. This material, together with site plans of the hull still remaining on the riverbed, may be displayed in South Australia's new Maritime Museum to be opened at Port Adelaide in 1986.[18]

It is very important that some priority be given to the study of Australian-built vessels. Our studies of Dutch, English and American wrecks are not reciprocated in those countries by studies of Australian-built wrecks. That task is up to us alone.

4. Historic Steamship Wrecks and Professional Salvors—New South Wales

It is a sad irony that New South Wales, the most populous State of Australia and the State with the largest number of shipwrecks along its coast, is the one Australian State which does not have an institutional programme of maritime archaeology.

The Maritime Archaeology Association of New South Wales was formed at a seminar in Sydney in 1978, the inaugural president being Associate Professor John Bach, a maritime historian at the University of Newcastle. The Association immediately prepared a submission to the State Premier to have the Commonwealth Historic Shipwrecks Act proclaimed in New South Wales. The legislation was proclaimed the following year, but to date the State Government has not followed up with any of its own initiatives, such as, for example, providing an officer to administer the Act in New South Wales. The Museum of Applied Arts and Sciences (Powerhouse Museum) was interested in maritime archaeology, but State funds were not forthcoming.

The Association's interest focussed on the wreck of a 299-ton paddle steamer, the *Ballina*, lost at the mouth of the Hastings River at Port Macquarie in 1879. Changing water currents had exposed about two-thirds of the wreck when it was found by a Department of Public Works diver doing survey work for a sea-wall extension. Surviving sections of paddle wheels and engines were noticed. As the site (protected under the Act) was seen as being under threat, a survey and recording expedition by Association members was given the approval of the Commonwealth Minister for Arts, Heritage and Environment. Members also carried out a survey and assessment of the small iron passenger-carrying steamer *Royal Shepherd*, wrecked in 1890. Like the *Xantho* wrecked off Western Australia, the *Royal Shepherd* had engines to the design of John Penn.[19]

Problems began to develop within the Maritime Archaeology Association of New South Wales as it became clear that for the near future at least the State was not willing to provide for a maritime archaeologist to be placed in an appropriate institution, and consequently that there could be no delegation of the responsibility for administering the Act. Some members feared that in the absence of professional direction the Association was catering for divers concerned with salvage for personal gain rather than those who wanted to further the aims of archaeology. So the Association has gradually dissolved. A new group, the Underwater Archaeological Research Group, was formed in Sydney with membership by invitation. Several regional groups also emerged: the Hastings Valley Maritime Archaeology Association at Port Macquarie in 1980, and the Maritime Archaeological Society of Newcastle in 1983.

The Underwater Archaeological Research Group, with representation from several staff members of the Museum of Applied Arts and Sciences, aims to develop a broader awareness of the discipline of archaeology and of the necessity for protection and conservation of sites by governments and the general public. In 1983 several members (under the direction of archaeologist Michael Lorimer) commenced excavation of the stern section of the wreck of the 199-ton iron screw steamer *John Penn*, an Illawarra Steam Navigation Company vessel wrecked south of Bateman's Bay in 1879.[20] The site is protected under the Act. The excavation by water dredge exposed the engine room for recording. The Group is also assembling a register of above-water shipwrecks and derelict vessels and has compiled a register of steam shipwrecks.[21]

The Hastings Valley Maritime Archaeology Association concentrated its interest on the site of the *Ballina*. The Association hoped to recover the *Ballina*'s paddle wheel and engines (considered to be a hazard to navigation) to form a focal point for a proposed Port Macquarie maritime museum, but these ideas were not realised.

The Maritime Archaeological Society of Newcastle has also focussed its interests on steamships, and has links with the Newcastle Maritime Museum. Members of the Group and the Society have carried out background research and conducted site surveys on the iron paddle steamer *Mimosa* (located by steam enthusiast John Riley) which was wrecked in 1863 and is now protected under the Commonwealth Act,[22] and what appears to be the paddle steamer *Commodore*, wrecked in 1931.[23] Since the survey was conducted on the latter site, it has sustained considerable damage from the dragging anchors of bulk carriers.

It is clear that some of these regional amateur groups, in carrying out basic research and survey projects, are performing a useful role. These groups would like to see a State Government representative appointed to administer the Act. In the absence of such a representative, responsible individuals and groups are inevitably frustrated and alienated. The result is a continuing massive destruction of the archaeological resource of New South Wales. Perhaps when the National Maritime Museum is established in Sydney some steps will finally be taken by the State Government to protect what remains of the underwater maritime heritage of New South Wales.

5. Maritime Archaeology in the Australian Territories

The Historic Shipwrecks Act automatically applies to territories administered by the Commonwealth Government. These include the Northern Territory, Norfolk Island, Heard Island, Macquarie Island, the Australian Antarctic Territory, the Cocos (Keeling) Islands, Christmas Island, and Ashmore and Cartier Shoals, which are administered through Canberra and the Northern Territory.

Although individual divers from each of the settled territories have shown interest in their local shipwreck sites, there are as yet no maritime archaeological associations. As it remains the normal practice that each site must be individually declared by the Minister to gain full protection under the Act, it is necessary for concerned divers to let the Commonwealth Government know the details of historic shipwreck sites in their areas.

The Department of Arts, Heritage and Environment has taken a number of initiatives in regard to sites in territorial waters. The Government declared as protected the Japanese submarine *I.124*, sunk by the Australian Navy off Darwin during World War II and now a war grave, after a local diver claimed to have attempted to destroy it with explosives. The diver is quoted in the *Australia Post* as saying:

> I frankly planned to make a shilling out of it. In fact, I expected that this wreck was going to make me a millionaire .... I threatened that if the wreck was going to be taken from me, then I would blow it up. .... I placed explosive charges and set them off, expecting that this would detonate the 42 mines which are on the sub's deck.[24]

Turbulence clouds the seabed with bubbles as a diver swims past an anchor on the *Sirius* wreck. (Photo: Pat Baker)

Another naval vessel, the 3,600-ton German light cruiser *Emden*, sunk at Cocos in 1914 by the Australian Navy, has also been declared protected under the Act.

An Australian Bicentennial Authority Project, organised by the Department of Arts, Heritage and Environment, involves investigating the remains of HMS *Sirius*, the principal warship of the First Fleet, lost in 1790 at Norfolk Island. I was asked to look at the site in December 1983, and to lead a 12-man expedition to the site in February 1985.[25] Local divers showed us the location of three anchors and a carronade lying under the outer edge of the surf zone, several hundred metres east of the landing pier at historic Kingston. Around these anchors were large rectangular iron ballast blocks, water-washed pebble ballast, rudder fittings and exterior hull fragments. It soon became clear that this material represented the position where the *Sirius* had first struck, and remained for 9 days. Then the ballast had fallen out through the bottom of the ship, and the hull was thrown more than her own length closer to shore.

Rudder chains from the *Sirius* are sketched by archaeologist Myra Stanbury. (Photo: Pat Baker)

Conservator Dr Ian MacLeod uses lamp black as a restoration treatment on a cannon reputedly from HMS *Sirius*. (Photo: Pat Baker)

One of the aims of the Bicentennial Project is to gain a better understanding of what happened to the ship and its contents after it struck the reef on 19th March 1790. Numerous artefacts have been seen further inshore, but it is known that several ships were wrecked off Kingston in later years. Other items, in government and private collections on Norfolk Island, are held by tradition to have come from the *Sirius*, but it requires some investigation to confirm this in each case. A final expedition, scheduled for February 1987, will attempt to confirm the location of the *Sirius*'s final resting place.

The records indicate that many shipwreck sites lie on or near the other Australian territories. These are now being found by divers at an increasing rate. Comparatively large numbers of itinerants work in these areas, and this increases the threat of cultural material being taken away, while at the same time reducing the viability of concerned amateur archaeological groups who might prevent the destruction of sites. These areas generally have less access to government expertise and funding for heritage protection than more densely populated mainland centres, so many of the sites, once located, are under some threat.

Chapter 9 REFERENCES

1. Atherton and Lester, 1982, p. 4.
2. Cook, 1983, pp. 38–41.
3. Henderson, unpublished 1978; and Henderson, 1980(4), pp. 31–35.
4. Atherton and Lester, p. 16.
5. Henderson, 1981(2); and Baker and Henderson, 1979.
6. Cumpston, 1964.
7. Grace, 1985, p. 17.
8. Steven, 1965, p. 23; and Hainsworth, 1971, p. 129.
9. Proudfoot, 1985, p. 21.
10. Carroll, 1982, p. 122.
11. Carroll *et al.*, 1984.
12. Schiffer and Gumerman, 1977. See also Coutts, 1982.
13. Staniforth, 1985.
14. Staniforth and Vickery, 1984, p. 6.
15. Staniforth (in press).
16. Marfleet, 1980, p. 21.
17. Jeffery, 1983, p. 88.
18. Jeffery, 1985.
19. Richards, 1980, p. 13.
20. Lorimer, 1985, pp. 22–23.
21. Lorimer, 1982, p. 84.
22. Riley, 1984.
23. Walters *et al.*, 1984.
24. Mawbey, 1978.
25. Henderson, 1985, pp. 44–46.

10 A Reflective Overview and Prospects for the Future

The field of maritime archaeology in Australia can boast very substantial advances since its beginning in the 1960s. The years ahead hold even more promise. But if we are to make the most of the challenge of the future we must pause and reflect upon our direction.

What have we Australians achieved? Most importantly the archaeologically significant shipwreck sites have been protected as cultural resources under comprehensive legislation. By this means the various governments (all Australian States and territories are now covered) have ensured the survival into the future of a substantial part of the underwater cultural heritage. This legislation pre-dates any protective measures provided by almost all other countries. It also substantially pre-dates any comprehensive protective measures by Australian governments for our terrestrial cultural resources. The process of legal control of the underwater cultural heritage was begun by the West Australian Government in 1964, and was widened dramatically in scope by the Commonwealth legislation of 1976. Since then the Victorian and South Australian Governments have plugged the gaps in their inshore sections of the protective net with complementary legislation. The recent prosecution in South Australia shows that the legislation has teeth.

We have endeavoured to ascertain the extent of our underwater cultural resources by establishing sites' registers in each State, and by implementing programmes of site inspection in most States. Documentation of the resource in Western Australia is now at a level probably unrivalled by any other region in the world. Archival work has yielded comprehensive information on thousands of shipwrecks, while the inspections have given accurate locations and measurements of several hundred sites. The Commonwealth is now supporting the endeavours of the various States in these aspects of site management. Over a hundred are protected under Commonwealth legislation, and the various State Acts substantially increase this tally. No other continent is as well organised in the assessment of submerged sites.

Departments of maritime archaeology have been established in government institutions in most States, and the West Australian sites initially seen as being under greatest threat from looters have been excavated according to archaeological principles, thus ensuring research results and preservation of the items raised. The Commonwealth Government has played a major role in stimulating and supporting particular excavation projects. Work on the *Pandora*, for example, although under the control of the State Government, has been principally a Commonwealth Government initiative, the Department of Arts, Heritage and Environment contributing $60,000 to the 1983 excavation season, and $70,000 to the 1984 season. The work on the wreck of HMS *Sirius* is wholly funded by the Australian Bicentennial Authority. The Commonwealth is also ensuring that substantial funds go to all States involved in maritime archaeology. It is unfortunate that several of the State Governments have not as yet followed up on the Commonwealth's far-sighted initiatives with consolidating work of their own.

In most cases (but not all), appropriate conservation facilities have been made available to the emerging maritime archaeology units. The conservation laboratory in Fremantle provides not only a facility for the processing on a regular basis of bulky and varied collections of waterlogged material but also produces excellent research results on the behaviour and characteristics of the various materials passing through the laboratory. These results are applicable at home and abroad, both in maritime archaeology and in wider fields.

The relics and information resulting from excavations have been used in a variety of ways to maximise their effectiveness, including displays for maritime museums, publications of both academic and popular nature, and information releases to the media. Shipwreck material is displayed in a variety of museums throughout the country, including the Western Australian Maritime Museum, the Fremantle Museum, the Albany Residency Museum, the Geraldton Museum, the Queensland Maritime Museum, the Queensland Museum, Ben Cropp's Shipwreck Museum, the Queen Victoria Museum and Art Gallery, the Flagstaff Hill Maritime Village at Warrnambool, and the South Australian Museum. More will be displayed in the South Australian Maritime Museum and the National Maritime Museum. The high public profile of maritime archaeology has benefitted maritime museums and all the other groups interested in maritime history.

As more States have accepted the responsibility and challenge of maritime archaeology some diversity has emerged in the structures and approaches of the responsible organisations. The diversity is to be welcomed but there are potential difficulties which can arise when field programmes are organised by one institution, conservation by another, curatorial responsibility by another and display perhaps in a fourth organisation. Such is the case, in varying degrees, in Victoria, South Australia and Tasmania. It could, of course, be argued that in institutions combining all of these facilities there may be a strong temptation to give priority to restoration and display even at the expense of the research ideals.

We have been active in the development of special on-site equipment and techniques, such as, for example, our three-dimensional grid system and experimentation in photogrammetry. Australia led the world in showing that it was possible to take a professional archaeological approach to turbulent shallow-water shipwreck sites.

We have now a postgraduate diploma course (based at the Western Australian Institute of Technology) to provide personnel (with a high degree of technical expertise) for public archaeology in most Australian States. This course, run in July 1980 and July 1981, is being repeated in 1986. We have also commenced input to university courses, with a prehistory honours unit at the University of Western Australia in 1985. An exciting new development is the first year archaeology course, planned to start at the University of Western Australia in 1987. This unit will include maritime archaeology, historical archaeology and prehistory. Thus, for the first time, students will be given a grounding in maritime archaeology at the commencement of a degree in archaeology.

The diving community of Australia has always been interested in shipwrecks, but in the past has shown little tolerance for the more passive aspects of site management in public archaeology. We have fostered a sympathetic and informed stance within the maritime archaeology associations towards conserving the underwater cultural resource, and these attitudes are influencing the general diving community.

Two international conferences have been held during the past decade to disseminate information and promote communication. In the future it would be useful to hold annual joint conferences (equal representation) with prehistory and with historical archaeology, to promote mutual awareness. With the creation of the Australian Institute for Maritime Archaeology (Inc.) in 1982, we have a medium for closer communication between practitioners in the various States. The Institute publishes a newsletter and regular bulletins, and has recently commenced a series of special publications. It has also assisted with archaeological fieldwork in Asian waters.

We have added to the volume of knowledge in a number of ways. The arguments cited in favour of pre-Dutch discovery of Australia have been substantially reduced by laboratory analysis of various artefacts from the sea. Examples include the guns and jars from the so-called Portuguese wreck at Cottesloe Beach, near Fremantle,[1] and the so-called Portuguese guns from Napier Broome Bay.[2]

The Dutch shipwrecks have yielded useful information about seventeenth century shipbuilding, maritime trade, warfare at sea, and navigational methods. The large collections from these accurately dated sites can be used to check upon established typologies. Some of the early post-settlement sites on remote sections of the Australian coastline are giving us a better understanding of trade routes and fishing activities taking place off our shores at that time. Little has yet been published on the later sites of the nineteenth century, but it is already clear that the well-preserved cargoes constitute a unique source of information about colonial life.

In the foreseeable future we can expect to see the results of the intensive fieldwork programmes in Western Australia during the 1970s, in the form of full archaeological reports on the excavations of the Dutch ships *Batavia* and *Vergulde Draeck*, and of the nineteenth century traders *Eglinton*, *James Matthews* and *Rapid*. This will result in the accumulation of a substantial data base within Australia for quantitative studies and pattern recognition. It will be possible to compare and contrast aspects of the collections from Dutch wrecks in West Australian waters with those of a similar period in European waters. Similarly, when more of the post-settlement sites in Australian waters have been excavated, classified and published, generalisations and predictions about the ships and crews should be possible.

Already some lines of enquiry are being pursued which deal quantitatively with collections and information acquired in part from underwater sites, examples being Green's catalogue of naval guns in Australia, and my study of the Australian pearling lugger.[3] The regional sites' registers have been developed very substantially in several States, and a computerised national sites' register is being formulated. Now that sites' registers have been developed it is more feasible for maritime archaeologists to seek the most appropriate sites to provide answers to questions they are interested in, rather than to devise questions about a particular site because they have already decided to excavate it. Examples of this are the current investigations of whaling ships and of India trade vessels.

With maritime archaeologists operating in most States it will be possible to explore on a national basis, hopefully with substantial help from national computer links, aspects of such themes as immigration, trade patterns, and Australian shipbuilding. When such programmes are put into effect it may be possible to draw more synthesis from the hotch-potch that comprises Australian maritime archaeology. If the proposed National Maritime Museum includes a substantial maritime archaeological research facility then it should go a long way towards providing such broader viewpoints.

A feature of maritime archaeology as it has developed in Australia is its strength at grass-roots level. Reflecting on the genesis of maritime archaeology in each of the States, a model of development emerges. In all States, amateur diving groups have provided focal points for diving enthusiasts interested in the preservation of those shipwreck sites which they feel are important to their history. This concern for protection and management generally involves members in on-site surveys to locate and describe the sites. Members do not normally become archaeologists in the accepted sense of the word (university trained), but they gain experience of many of the on-site techniques practised by archaeologists.

The survey work generally heightens the awareness of group members to the extent of damage caused by fossickers on historic shipwreck sites, and the members become a pressure group, encouraging governments to adopt and administer protective legislation, calling for a government institution to provide a display venue for relics, and seeking professional advice from such an institution as to how members might best direct their energies. This advice would need to cover such issues as which sites might be regarded as historically or archaeologically significant, and what the aims and procedures on a particular shipwreck survey should be.

If the State government is receptive to the idea of protective legislation, and a government institution is willing to take the lead, then the collective skills and energy of the group can be harnessed for assistance to the institutional archaeologists who direct any excavation work and carry out research. The institution must have a permanent involvement for adequate management of the sites, proper maintenance of any collections and on-going study of the material.

Maritime archaeologists have reciprocated by welcoming amateur involvement in their fieldwork and laboratory activities. But maritime archaeology's contacts and credentials on the academic side are less developed. There are no maritime archaeologists holding permanent (or temporary) positions in Australian universities. The archaeology components of the Western Australian Institute of Technology courses are taught by museum (public) archaeologists rather than university (academic) archaeologists, and only one maritime archaeologist has completed a higher degree in archaeology at an Australian university (several of us have taken what appeared to be the closest alternative and completed postgraduate studies in history departments, but the general swing within historical archaeology away from the discipline of history and towards that of anthropology now makes such a course less useful.

This unsatisfactory situation is in marked contrast with the positions of prehistory and terrestrial historical archaeology in Australia. To a lesser degree it also contrasts with the position of maritime archaeology in some overseas countries, masters programmes being available in the United States, Israel, Wales and Scotland, for example.

The past 20 years have been characterised by radical methodological and theoretical changes in mainstream archaeology, the new emphasis, after the turmoil, being on culture process rather than on culture history. Prehistory and historical archaeology in Australia have very consciously moved towards the new approach. Arguments can be made for and against the idea of Australian maritime archaeology joining this movement, but so far the arguments have not been adequately considered. My own view is that unless a significant number of the practitioners of Australian maritime archaeology make the change then this field of study will have its credibility reduced, both within the field of maritime archaeology overseas and within mainstream archaeology in Australia. We must, while avoiding unnecessarily destructive self-criticism, recognise the need for a theoretical perspective within maritime archaeology, to more fully comply with contemporary practice.

To effectively develop that awareness it is necessary that we have university representation—an academic dimension. Placement of a practitioner in an academic position is a necessary step if Australian maritime archaeology is to fully progress from pioneer status to maturity. Such representation would also be good for prehistory and historical archaeology in Australia, whose practitioners have, since John Mulvaney's one paragraph back in 1975, ignored the potential and contributions of maritime archaeology (it does not rate a mention in *Australian Field Archaeology* or other recent titles in Australian archaeology).[5] One example of the potential benefit to prehistorians of closer communication is that of site locating. A major question question for prehistorians relates to whether the Aborigines of the past spent most of their time living on the coast (areas now inundated by up to 100 metres of water). Only the examination of numbers of underwater sites can answer that question. Closer communication with experienced underwater archaeologists might be expected to assist in this search.

The academic goals within maritime archaeology should not be pursued at the expense of weakening the important links with amateurs. For some time now, diving groups have been calling for undergraduate or school courses, to make the area of study more accessible. Even diver training courses are now including aspects of the historical backgrounds of shipwrecks. For the growing army of qualified divers to adopt responsible, appreciative attitudes towards archaeologically significant sites, it is necessary for the profession to have some input to these courses also.

Easily accessible courses are a way for maritime archaeologists to influence the activities of the general diving community on wreck sites. Now that professional maritime archaeologists are operating in almost all States of Australia, it is an appropriate time for them to review their own activities on shipwreck sites. In our isolation from academic archaeology, and to a lesser extent from the prehistory and historical archaeology professions, we have not been sufficiently involved in questions of professional ethics. Until very recently neither the *Burra Charter* nor the *Code of Ethics of the Australian Association of Consulting Archaeologists* have been the subject of regular discussion among Australian maritime archaeologists. In past years it might have been arguable that there was such a small group of maritime archaeologists, all operating from permanent positions within government institutions, that a code of ethics was not necessary. Now, with short-term consultancies, and practitioners throughout Australia, there is an urgent need for a code of ethics.

In mainstream archaeology today there is a growing sense of the need for parsimonious use of resources. Should maritime archaeologists in Australia cease all excavations, or be excavating fewer sites, and smaller sections of them, in the future? The high public profile of maritime archaeology has undoubtedly given prehistorians and historical archaeologists the impression that we are gung-ho, to say the least, in our readiness to embark upon excavation of sites. The reality is that only a small number of maritime archaeological sites have been excavated in Australian waters. In only one case, that of the *Batavia* excavation, can it be claimed that anything approaching total excavation has taken place. No other hulls (and hull remains generally comprise the greatest part of the more complete sites) have been raised by maritime archaeologists from Australian waters. It is true that the Western Australian Museum has a large collection of artefacts, principally from the *Batavia*, and that some of those artefacts remain in storage awaiting treatment. But the standard of care, from what is the most sophisticated laboratory of its kind in the Southern Hemisphere, compares extremely well with the treatment given to collections made by prehistorians and historical archaeologists.

We have not been profligate in the past. The excavation programme in Western Australia was commenced after it had been clearly demonstrated that a system of wardens alone was not working. Today the legislation and its policing are more effective, but there had not been a successful prosecution under any of these Acts until last year (1985), and most of the sites adjacent to the heaviest population centres (and hence most subjected to interference) are not under any regular police surveillance.[6] Diver attitudes have changed very substantially, but the numbers of divers visiting sites has accelerated drastically, and a very small minority of looters is sufficient to destroy sites. In the short term, public education programmes can reduce, but not entirely overcome, this problem. Thus the need continues for an active excavation programme.

The nature of the maritime archaeological resource in Australia lends itself to multicultural studies. The initial Eurocentric bias is now being balanced by conscious efforts in some areas to find and study Australian-built vessels, and the results may be expected to reflect a more mature approach to Australia's past. Maritime archaeology is an appropriate vehicle for the exploration of Australians' cultural identity.

Chapter 10   REFERENCES

1. Henderson, 1973.
2. Green, 1982, pp. 73-83.
3. Henderson, 1981, pp. 35-38. The lugger study was funded by the Australian Research Grants Scheme. The guns catalogue was funded by the Australian War Memorial Museum.
4. Masters and Flemming, 1983.
5. Mulvaney, 1975; Connah, 1983. Bowdler, 1985, is the first exception.
6. McKinlay and Henderson, 1985.

APPENDIX

Shipwrecks Declared Under Legislation

A. The Commonwealth's Historic Shipwrecks Act, 1976

State or Territory	Name by which Ship Known	Type of Ship or Remains	Date of Wreck	Location of Remains Latitude	Location of Remains Longitude	Date Declared Historic	Protected Zone
NSW	*Ballina*	paddle steamship	1879	31°25.5′S	152°54.9′E	11.4.79	
	John Penn	250-ton twin-screw steamship	1879	35°51.1′S	150°11.3′E	29.10.82	
	Mimosa	153-ton iron paddle steamship	1863	36°34.9′S	150°03.7′E	29.10.84	
	Satara	5,272-ton screw steamship	1910	32°28.8′S	152°31.7′E	16.11.84	22.11.84
	Nobbys Head Wreck	paddle steam tug	1931?	32°55.4′S	151°52.3′E	29.10.84	
QLD	HMS *Pandora*	British naval vessel	1791	11°22′S	143°58′E	18.11.77	5.6.81
	Yongala	screw steamship	1911	19°18.3′S	147°37.3′E	5.6.81	14.1.83
	Aarhus	640-ton sailing barque	1894	26°59.7′S	153°28.7′E	22.10.81	19.12.83
	Gothenburg	700-ton barque-rigged steamship	1875	19°21.2′S	148°2.5′E	22.10.81	
	Morning Star	140-ton brig	1814	12°25.8′S	143°25.3′E	22.10.81	
	Mermaid	84-ton schooner	1829	17°5.8′S	146°11.5′E	22.10.81	
	Quetta	3,302-ton steamship	1890	10°39.9′S	142°37.7′E	22.10.81	
	Foam	wooden labour vessel	1893	18°16.3′S	147°23.2′E	14.1.83	Yes
	HMS *Endeavour*	British naval vessel material jettisoned	1770			19.12.83	
	Scottish Prince	950-ton iron paddle steam tug	1887	27°57.5′S	153°26.2′E	29.10.84	
SA	*Loch Vennachar*	1,552-ton iron clipper ship	1905	35°53′S	136°31.9′E	12.3.82	
	Montebello	2,284-ton iron barque	1906	36°01.3′S	137°00.2′E	12.3.82	
	Clan Ranald	3,596-ton turret-deck screw steamship	1909	35°10.1′S	137°37.5′E	12.3.82	
	Geltwood	1,056-ton iron barque	1876	37°37.2′S	140°10.1′E	25.2.83	
	Admella	360-ton iron single-screw steamship	1859	37°52.7′S	140°21.0′E	13.7.83	
	Margaret Brock	245-ton 3-masted wooden barque	1852	36°57.0′S	139°35.7′E	13.7.83	
TAS.	*Sydney Cove*	3-masted wooden vessel	1797	40°29.6′S	148°04.7′E	16.3.84	
	City of Edinburgh	367-ton wooden barque	1840	40°01.3′S	147°53.7′E	1.10.85	
	Litherland	305-ton wooden ship	1853	40°33.0′S	148°07.0′E	1.10.85	
	Cambridgeshire	1,691-ton iron ship	1875	40°28.6′S	148°01.8′E	1.10.85	
	Asterope	602-ton wooden barque	1883	41°02.9′S	146°44.2′E	1.10.85	

State or Territory	Name by which Ship Known	Type of Ship or Remains	Date of Wreck	Location of Remains Latitude	Location of Remains Longitude	Date Declared Historic	Protected Zone
NT	I.124	Japanese submarine	1942	12° 50.5' S	130° 06.8' E	12.7.77	12.7.77
COCOS	Emden	3,600-ton light cruiser	1914	11° 50.5' S	96° 49.4' E	12.3.82	12.3.82
NORFOLK	HMS Sirius	British naval vessel	1790	29° 03.7' S	167° 57.1' E	29.10.84	
VIC.	Children	786-ton steamship sailing barque	1839	38° 29.5' S	142° 40.4' E	11.3.82	
	Monumental City	wooden clipper	1853	37° 33.5' S	149° 50.7' E	11.3.82	
	Schomberg	iron sailing ship	1855	38° 37' S	152° 53.3' E	11.3.82	
	Loch Ard	64-ton wooden schooner	1878	38° 39.1' S	143° 03.5' E	11.3.82	
	Thistle	297-ton paddle steamship	1836	38° 23.1' S	142° 14.5' E	16.11.84	
	Clonmel	405-ton barquentine	1841	38° 40' S	146° 42' E		
	La Bella	Dutch East Indiaman	1905	38° 24.4' S	142° 29.2' E	23.4.82	
WA	Batavia		1629	Wallabi Group, Houtman Abrolhos		15.12.76	
	Vergulde Draeck or Gilt Dragon	Dutch East Indiaman	1656	Ledge Point		15.12.76	
	Zuytdorp	Dutch East Indiaman	1712	Zuytdorp Cliffs		15.12.76	28.11.78
	Zeewijk	Dutch East Indiaman	1727	Pelsaert Group, Houtman Abrolhos		15.12.76	
	Trial	English East Indiaman	1622	20° 17.5' S	115° 21.1' E	8.9.77	
	James	200-ton wooden brig	1830	32° 05.1' S	115.° 45.0' E	8.9.77	
	Lancier	British barque	1839	32° 04.6' S	115° 38.1' E	8.9.77	
	Elizabeth	194-ton barque	1839	32° 00.8' S	115° 44.9' E	8.9.77	
	Ocean Queen	268-ton barque	1840	28° 56.5' S	113° 51.7' E	8.9.77	
	James Matthews	167-ton snow brig ex-slaver	1841	32° 08.0' S	115° 44.5' E	8.9.77	
	Cervantes	American whaling barque	1844	30° 30.7' S	115° 02.0' E	8.9.77	
	Arpenteur	schooner	1849	34° 50.0' S	118° 24.0' E	8.9.77	
	Eglinton	464-ton barque	1852	31° 39.0' S	115° 40.0' E	8.9.77	
	Lady Lyttleton	wooden barque	1867	34° 59.9' S	117° 56.8' E	8.9.77	
	Ben Ledi	1,000-ton iron ship	1879	28° 55.5' S	113° 59.6' E	8.9.77	
	Centaur	118-ton iron brig	1874	31° 51.7' S	115° 42.8' E	8.9.77	
	Chalmers	606-ton wooden barque	1874	32° 22.0' S	115° 41.2' E	8.9.77	
	Fairy Queen	pearling schooner	1875	21° 49.2' S	113° 11.5' E	8.9.77	
	Zedora	269-ton barque	1875	32° 04.1' S	115° 37.7' E	8.9.77	

State or Territory	Name by which Ship Known	Type of Ship or Remains	Date of Wreck	Location of Remains Latitude	Location of Remains Longitude	Date Declared Historic	Protected Zone
WA	Gem	52-ton colonial cutter	1876	31°59.6'S	115°33.5'E	8.9.77	
	Hero of the Nile	356-ton barque	1876	32°23.2'S	115°43.5'E	8.9.77	
	Georgette	332-ton steamship	1876	34°02.0'S	114°59.4'E	8.9.77	
	Lady Elizabeth	658-ton barque	1878	32°01.1'S	115°32.8'E	8.9.77	
	James Service	441-ton iron barque	1878	32°27.5'S	115°39.5'E	8.9.77	
	Marten	28-ton schooner	1878	28°55.5'S	113°59.6'E	8.9.77	
	Diana	224-ton 3-masted schooner	1878	32°05.9'S	115°45.3'E	8.9.77	
	Star	69-ton schooner	1880	32°22.2'S	115°41.2'E	8.9.77	
	Agincourt	443-ton barque	1882	34°12.1'S	115°01.18'E	8.9.77	
	Macedon	876-ton steamship	1883	31°59.3'S	115°33.3'E	8.9.77	
	Chaudiere	470-ton barque	1883	34°12.2'S	115°01.7'E	8.9.77	
	Mira Flores	550-ton iron barque	1886	32°00.3'S	115°27.9'E	8.9.77	
	Belle of Bunbury	42-ton schooner	1886	32°41.4'S	115°41.4'E	8.9.77	
	Janet	211-ton schooner	1887	31°59.3'S	115°33.3'E	8.9.77	
	Denton Holme	998-ton iron barque	1890	31°59.3'S	115°33.3'E	8.9.77	
	Day Dawn	500-ton wooden barque	1890	32°13.9'S	115°41.2'E	8.9.77	
	Raven	362-ton wooden barque	1891	32°01.3'S	115°33.0'E	8.9.77	
	Dato	200-ton wooden brig	1893	32°14.3'S	115°41.4'E	8.9.77	
	Ulidia	1,378-ton iron barque	1893	32°03.3'S	115°37.7'E	8.9.77	
	Omeo	789-ton barque-rigged iron steamship	1894	32°06.4'S	115°45.6'E	8.9.77	
	Sepia	725-ton iron barque	1898	32°08.0'S	115°38.4'E	8.9.77	
	City of York	1,167-ton iron barque	1899	31°59.7'S	115°29.1'E	8.9.77	
	Carlisle Castle	1,484-ton iron barque	1899	31°20.0'S	115°37.9'E	8.9.77	
	Rapid	366-ton wooden ship	1811	22°44'S	113°41'E	15.12.78	
	Europa	iron barque	1897	30°24.9'S	114°59.5'E	13.6.79	
	Eyre Wreck	scattered wreckage indicating a substantial vessel	—1830	31°18'S	126°51'E	21.7.81	
	Sunset Beach Wreck	wreckage of a 300–400-ton vessel	—1870	28°43.4'S	114°37'E	21.7.81	
	Manfred	587-ton barque	1879	16°1.4'S	122°07.7'E	21.7.81	
	Browse Island Wreck	large iron sailing vessel of about 1,000 tons	1870s or 1880s	14°07'S	123°33.6'E	21.7.81	
	Perth	499-ton iron coastal steamship	1887	22°41.4'S	113°38.5'E	21.7.81	

State or Territory	Name by which Ship Known	Type of Ship or Remains	Date of Wreck	Location of Remains Latitude	Location of Remains Longitude	Date Declared Historic	Protected Zone
WA	Zuir	iron steamship	1902	22° 36.1' S	113° 37.1' E	21.7.81	
	Eddystone	2,040-ton brigantine-rigged iron steamship	1894	20° 36.4' S	117° 44.0' E	21.7.81	
	Villalta	866-ton steel barque	1897	31° 19' S	115° 27.0' E	21.7.81	
	Katinka	iron barque	1900	34° 12' S	115° 2' E	21.7.81	
	Karrakatta	1,271-ton schooner-rigged coastal steamship	1901	16° 21.4' S	123° 02.1' E	21.7.81	
	Crown of England	1,847-ton sailing ship	1912	20° 37.1' S	117° 43.9' E	21.7.81	
	Fin	iron whale catcher	1923	22° 39' S	113° 38.0' E	21.7.81	
	Windsor	iron steamship	1908	29° 0.0' S	113° 56.30' E	21.7.81	
	Mayhill	4-masted barque	1895	28° 45.9' S	114° 34.2' E	21.7.81	
	Ringbolt Bay Wreck	Wreckage suggesting small vessel	–1880	34° 21.3' S	115° 9.2' E	21.7.81	
	Hadda	334-ton wooden barque	1877	28° 28.4' S	113° 47.5' E	12.3.82	
	Mayflower	277-ton brig	1880	34° 20' S	115° 10.5' E	12.3.82	
	Uribes	118-ton iron schooner	1942	32° 0.2' S	115° 33.3' E	12.3.82	
	Rowley Shoals Wreck	armed whaler of 200–250 tons	–1810	17° 05.4' S	119° 35.5' E	12.3.82	
	Priestman Dredge	grab dredge	1893	32° 03.5' S	115° 44.1' E	14.1.83	
	Contest	wooden barque	1874	32° 16.5' S	115° 42.8' E	14.1.83	
	Koombana	screw steamship	1912	19° 05' S	118° 52' E	14.1.83	
	Batoe Bassi	400-ton wooden barque	1880	33° 54.6' S	122° 50.2' E	14.1.83	
	Cumberland	44-ton wooden ship	1830	34° 17.5' S	115° 02.3' E	26.5.83	
	Lubra	321-ton iron steamship	1898	30° 18.3' S	114° 59.7' E	19.12.83	
	Highland Forest	998-ton iron barque	1901	115° 40' S	32° 23' E	1986	
	Yarra	482-ton iron barque	1884	14° 02.3' S	121° 46.0' E	1986	

B. South Australian Historic Shipwrecks Act, 1981

Name by which Ship Known	Type of Ship or Remains	Date Wrecked	Location of Remains Latitude	Location of Remains Longitude	Date Shipwreck or Relics Declared Historic	Protected Zone
Star of Greece	1,227-ton 3-masted iron ship	1888	35° 15.2 ' S	138° 27.5 ' E	18.2.82	
Grecian	518-ton 3-masted wooden barque	1850	34° 47.5 ' S	138° 28 ' E	18.2.82	
Nashwauk	762-ton 3-masted wooden ship	1855	35° 12 ' S	138° 28 ' E	18.2.82	
Santiago	455-ton 3-masted iron barque	abandoned 1945	34° 48.7 ' S	138° 32.4 ' E	18.2.82	
Tigress	225-ton 3-masted wooden snow brig	1848	35° 11.2 ' S	138° 27.8 ' E	18.2.82	
Norma	2,122-ton 4-masted steel barque	1907	34° 49.6 ' S	138° 25.2 ' E	18.2.82	
Water Witch	wooden cutter	1842	34° 24.5 ' S	139° 36.9 ' E	14.4.83	
Zanoni	338-ton composite barque	1867	34° 30.8 ' S	138° 03.7 ' E	12.5.83	26.5.83
Marion	809-ton 3-masted wooden ship	1851	35° 09.7 ' S	137° 48.4 ' E	14.4.83	
Iron King	871-ton 3-masted iron ship	1873	35° 09.7 ' S	137° 48.7 ' E	14.4.83	
City of Adelaide	steel hydraulically propelled lifeboat	abandoned 1954	34° 44.6 ' S	135° 52.4 ' E	14.4.83	
San Miguel	535-ton 3-masted iron barque	1865	34° 04.4 ' S	137° 27.5 ' E	14.4.83	
Solway	337-ton 3-masted wooden ship	1837	35° 35.1 ' S	138° 35.9 ' E	26.5.83	

C. Victorian Historic Shipwrecks Act, 1982

Name by which Ship Known	Type of Ship or Remains	Date Wrecked	Location of Remains Latitude	Longitude	Date Shipwreck or Relics Declared Historic	Protected Zone
City of Launceston	368-ton screw steamship	1865	38° 04.7' S	144° 49.5' E	17.3.82	Yes. NO ACCESS
William Salthouse	251-ton wooden brig	1841	38° 16.6' S	144° 42.3' E	20.12.82	Yes. LIMITED ACCESS
Mountain Maid	160-ton wooden brig	1856	38° 14.4' S	144° 42.4' E	13.3.85	
Clarence	67-ton wooden schooner	1850	38° 12' S	144° 43' E	1985	Yes

D. West Australian Maritime Archaeology Act, 1973

Name by which Ship Known	Type of Ship or Remains	Date Wrecked	Location of Remains Latitude	Longitude
Sunbeam	72-ton iron auxiliary steam yacht	1892	14° 20.6' S	126° 01.8' E
Xantho	97-ton iron steamship	1872	28° 11.2' S	114° 14.1' E
Calliance	822-ton wooden ship	1864	15° 30.5' S	124° 36.7' E

Bibliography

A. Published Works

Agnew N.H., 1983: Maritime Archaeology in Queensland. In Amess, J., and Jeffery, W. (eds), *Proceedings of the Second Southern Hemisphere Conference on Maritime Archaeology* (Adelaide), pp. 109-16.

Allen, J., Golson, J., and Jones, R. (eds) 1977: *Sunda and Sahul: prehistoric studies in Southeast Asia, Melanesia and Australia* (Academic Press, London).

Amess, J., 1983: Operation of the Commonwealth Historic Shipwrecks Act. In Amess, J., and Jeffery, W. (eds), *Proceedings of the Second Southern Hemisphere Conference* on Maritime Archaeology (Adelaide), pp. 47-56.

Anderson, J.L., 1981: The Mahogany Ship: History and Legend *The Great Circle* 3(1), pp. 46-7.

Anon., 1982-1983: *Cootamundra News* 1-6 (Surrey, England).

Ariel, A., 1984: Navigating with Kenneth McIntyre: A Professional Critique. *The Great Circle*, 6(2) pp. 135-9.

Ashdown, John, 1972: Oil Jars. *The International Journal of Nautical Archaeology* 1, pp. 147-53.

Atherton, K., and Lester, S., 1982: *Sydney Cove* Site Work 1974-1980, An Overview. *Bulletin of Australian Institute for Maritime Archaeology* 6, p. 4.

Bach, John, 1976: *A Maritime History of Australia* (Nelson, Melbourne).

Baker, Patrick, and Henderson, Graeme, 1979: *James Matthews* excavation. A second interim report. *The International Journal of Nautical Archaeology* 8(3), pp. 225-44.

Barkman, L., 1977: Treatment of Waterlogged Finds. In Green, J. (ed.), *Papers from the First Southern Hemisphere Conference on Maritime Archaeology* (Oceans, Melbourne), pp. 120-6.

Bass, George, 1966: *Archaeology Underwater* (Thames and Hudson, London).

—— (ed.), 1972: *A History of Seafaring, Based on Underwater Archaeology* (Thames and Hudson, London).

——, 1980: Marine Archaeology: A Misunderstood Science. In Borgese, E.M., and Ginsburg, N., (eds). *Ocean Yearbook 2* (University of Chicago Press, Chicago), pp. 137-52.

Bateson, Charles, 1972: *Australian Shipwrecks. Volume 1: 1622-1850* (Reed, Sydney).

Blackman, H., 1976: *The Armouries of the Tower of London: Ordinance* (H.M. Stationery Office, London).

Blainey, Geoffrey, 1966: *The Tyranny of Distance. How Distance Shaped Australia's History* (Sun Books, Melbourne).

Bolton, G.C., 1977: ANCODS—Australia-Netherlands Committee on Old Dutch Shipwrecks. In Green, J. (ed.), *Papers from the First Southern Hemisphere Conference on Maritime Archaeology* (Oceans, Melbourne), pp. 28-30.

Bowdler, Sandra, 1976: Left High and Dry. *Hemisphere* 20(5), pp. 29-33.

——, 1977: The coastal colonisation of Australia. In Allen, J., Golson, J., and Jones, R. (eds), *Sunda and Sahul* (Academic Press, London), pp. 205-46.

———, 1985: *Revealing the Past. Archaeology in Australia* (Methuen, North Ryde, New South Wales).
Broeze, F.J.A., 1975: The Cost of Distance. Shipping and the early Australian economy, 1788-1850. *The Economic History Review* 28, pp. 582-97.
———, 1978: British Intercontinental Shipping and Australia, 1813-1850. *The Journal of Transport History* 4(4), pp. 189-207.
———, 1982: Private Enterprise and the Peopling of Australia, 1831-1850. *The Economic History Review* 35(2), pp. 235-53.
Bruijn, Jaap, 1980: Between Batavia and the Cape: Shipping Patterns of the Dutch East India Company. *Journal of Southeast Asian Studies* 11(2), pp. 251-61.
Bruijn, Jaap, and Van Eyck, Els, 1982: Mutiny: Rebellion on the Ships of the Dutch East India Company. *The Great Circle* 4(1), pp. 1-9.
Bugler, A.R., 1966: *HMS 'Victory', Building, Restoration and Repair* (H.M. Stationery Office, London).
Carroll, D., 1983. The Role of the Amateur in Maritime Archaeology. In Amess, J., and Jeffery, W. (eds). *Proceedings of the Second Southern Hemisphere Conference on Maritime Archaeology* (Adelaide).
Chapman, F.H., 1775: *Architectura Navalis Mercatoria* (Sweetman 1967 reprint, New York).
Commonwealth of Australia, 1914- : *Historical Records of Australia* (Library Committee of the Commonwealth Parliament, Sydney).
Connah, Graham, (ed.), 1983: *Australian Archaeology, A Guide to Techniques.* (Australian Institute of Aboriginal Studies, Canberra).
Cook, C., 1982: Masons Cove, Port Arthur, Maritime Archaeological Survey. *Bulletin of the Australian Institute for Maritime Archaeology* 7(2), pp. 38-41.
Cotsell, G., 1856: *A Treatise on Ship's Anchors* (John Weale, London).
Coutts, P., 1982: Management of the Aboriginal Cultural Heritage in Victoria. *Records of the Victoria Archaeological Survey* 13, pp. 85-114.
Crawford, I.M., 1968: *The Art of the Wandjina: Aboriginal cave paintings in Kimberley, Western Australia* (Oxford University Press, Melbourne).
———, 1977: Maritime Archaeology Legislation in Western Australia. In Green, J. (ed.), *Papers from the First Southern Hemisphere Conference on Maritime Archaeology* (Oceans, Melbourne), pp. 30-3.
Crowley, F.K., (ed.), 1974: *A New History of Australia* (Heineman, Melbourne).
Cumpston, J.S., 1964: *Shipping Arrivals and Departures at Sydney, 1788-1825* (The Author, Canberra).
Department of Home Affairs, 1980: *Directory of National Estate Studies, 1973-4 and 1979-80* (Australian Government Publishing Service, Canberra).
De Burgh, W.J., and Henderson, Graeme, 1979: *The Last Voyage of the 'James Matthews'* (Western Australian Museum, Perth).
Drake-Brockman, Henrietta, 1963: *Voyage to Disaster* (Angus and Robertson, Sydney).
Fairbridge, R., 1983: Oral presentation referred to in Masters, P., and Flemming, N. (eds), Quaternary Coastlines and Marine Archaeology (Academic Press, London).
Fairburn, William, 1945: *Merchant Sail* (Bath).
Falconer, William, 1780: *Falconers Marine Dictionary, 1780* (David and Charles reprint, Newton Abbot, 1970).
Flemming, N.C., 1985: Ice Ages and Human Occupation of the Continental Shelf. *Oceanus* 28(1), pp. 18-23.
Flood, Josephine, 1983: *Archaeology of the Dreamtime* (Collins, Sydney).
Florian, M.L., 1977: The Physical, Chemical and Morphological Conditions of Marine Archaeological Wood should dictate the Conservation Process. In Green, J. (ed.), *Papers from the First Southern Hemisphere Conference on Maritime Archaeology* (Oceans, Melbourne), pp. 128-44.

Gould, Richard (ed.), 1983: *Shipwreck Anthropology* (University of New Mexico Press, Albuquerque).
Grace, Roger, 1985: Duskey's Big Guns. *Aorangi* 4(1), pp. 15-18.
Green, J. N., 1973: The Wreck of the Dutch East Indiaman the *Vergulde Draeck*, 1656. *The International Journal of Nautical Archaeology* 2(2), pp. 267-90.
———, 1975: The VOC ship *Batavia* wrecked in 1629 on the Houtman Abrolhos, Western Australia. *The International Journal of Nautical Archaeology* 4(1), pp. 43-64.
———, 1977(1): *Australia's Oldest Wreck: The Loss of the 'Trial', 1622*. (British Archaeological Reports, Oxford).
———, 1977(2): The Western Australian Museum Maritime Archaeology Department and the Dutch Wreck Programme. In Green, J. (ed.), *Papers from the First Southern Hemisphere Conference on Maritime Archaeology* (Oceans, Melbourne), pp. 62-8.
——— (ed.), 1977(3): *The VOC Jacht 'Vergulde Draeck' Wrecked Western Australia 1656* (British Archaeological Reports, Oxford).
———, 1982: The Carronade Island Guns and Australia's Early Visitors. *The Great Circle* 4(2), pp. 73-83.
Grolier Society of Australia, 1965: *Australian Encyclopedia* (Sydney).
Hainsworth, D. R., 1971: *The Sydney Traders* (Cassell, Melbourne).
Hallam, S., 1983: The Peopling of the Australian Continent. *Indian Ocean Newsletter* 4(1), pp. 1-6.
Halls, C., 1965: The Loss of the Ridderschap van Holland. *Annual Dog Watch* 22, pp. 3-8.
Heeres, J. G., 1899: *The Part Borne by the Dutch in the Discovery of Australia 1606-1765* (Luzac for the Royal Dutch Geographical Society, London).
Henderson, Graeme, 1973: *The Wreck of the 'Elizabeth'. Studies in Historical Archaeology.* (Australian Society for Historical Archaeology, Sydney).
———, 1980(1): Finds from the wreck of HMS *Pandora*. *The International Journal of Nautical Archaeology* 9(3), pp. 237-42.
———, 1980(3): *Unfinished Voyages. Western Australian Shipwrecks 1622-1850* (University of Western Australia Press, Nedlands).
———, 1980(4): Three Post Australian Settlement Shipwreck Sites: HMS *Pandora* (1791), the *Sydney Cove* (1797) and an unidentified site near North West Cape. *The Great Circle* 2(1), pp. 24-41.
———, 1981(1): The Development of the Australian Pearling Lugger. *The Bulletin of the Australian Institute for Maritime Archaeology* 5, pp. 35-8.
———, 1981(2): The American China Trader *Rapid* (1811): An Early Australian Shipwreck Site Identified. *The Great Circle* 3(2), pp. 125-32.
———, 1983(1): The Excavation of a Shipwreck Site off Western Australia *Indian Ocean Newsletter* 4(2), pp. 3-4.
———, 1985: Raise the Remains of the *Sirius*? *Skindiving in Australia*, 15(1), pp. 44-6.
Henderson, Graeme, Lyon, David, and MacLeod, Ian, 1983(2): HMS *Pandora* Lost and Found. *Archaeology* 36(1), pp. 28-35.
Henderson, James, 1982: *Marooned* (St George, Perth).
Holthouse, Hector, 1976: *Ships in the Coral* (MacMillan, Melbourne).
Horsburgh, James, 1809: *Directions for Sailing to and from the East Indies.* (London).
Hough, Richard, 1972: *Captain Bligh and Mr Christian.* (Hutchinson, London).
Hundley, Paul, 1983: Batavia Reconstruction. In Amess, J., and Jeffery, W. (eds). *Proceedings of the Second Southern Hemisphere Conference on Maritime Archaeology* (Adelaide), pp. 249-59.
ICOMOS, 1981: *The Australian ICOMOS Charter for the Conservation of Places of Cultural Significance (The Burra Charter).*
Ingleman-Sundberg, C., 1977: The VOC ship *Zeewijk* lost off the Western Australian coast in 1727. An interim report on the first survey. *The International Journal of Nautical Archaeology* 6, pp. 225-32.

———, 1978: Relics from the Dutch East Indiaman *Zeewijk* (Western Australian Museum, Perth).
Jeffery, Bill, 1983: The Development of Maritime Archaeology in South Australia. In Amess, J., and Jeffery, W. (eds), *Proceedings of the Second Southern Hemisphere Conference on Maritime Archaeology* (Adelaide), pp. 83–92.
Jones, Rhys, and Meehan, Betty, 1977: Floating Bark and Hollow Trunks. *Hemisphere*, 21(4), pp. 16–21.
Loney, Jack, 1980: *Australian Shipwrecks, Volume 2: 1851–1870* (Reed, Sydney).
———, 1982: *Australian Shipwrecks, Volume 3: 1871–1900* (List Publications, Geelong).
Lorimer, Michael, 1980: Recent Maritime Archaeological Investigations in New South Wales. *The Australian Society for Historical Archaeology Newsletter* 10(3), pp. 27–38.
———, 1982: Underwater Archaeological Research Group *The Bulletin of the Australian Institute for Maritime Archaeology* 6, p. 84.
———, 1985: The John Penn Reappears. *Australian Sea Heritage*, No. 5, pp. 22–3.
McCarthy, Mike, 1979(1): The Excavation and Identification of the ex-American whaler *Day Dawn*. *The International Journal of Nautical Archaeology* 3, pp. 157–60.
——— (ed.), 1979(2): *MAAWA 1974–1978, A Review of the Past Four Years Involvement in Maritime Archaeology and History* (The Author, Fremantle).
———, 1984: Western Australia's First Coastal Steamer. *Australian Association for Maritime History Newsletter* No. 19, p. 5.
McGimsey, C. R., 1972: *Public Archaeology* (Seminar Press, New York).
McGrail, S., 1974: A 17th Century Gunners Tallystick. *The International Journal of Nautical Archaeology* 3, pp. 157–60.
MacGregor, D., 1973: *Fast Sailing Ships 1775–1875* (Nautical Publishing Co., Lymington, Hampshire).
McIntyre, Kenneth, 1977: *The Secret Discovery of Australia. Portuguese ventures 200 years before Captain Cook* (Souvenir Press, Menindie, South Australia).
McKinlay, J., and Henderson, G., 1985: The Protection of Historical Shipwrecks: A New Zealand Case Study. *Archaeology* 38(6), pp. 48–51.
Macknight, Campbell, 1969: *The Farthest Coast. A Selection of Writings Relating to the History of the Northern Coast of Australia* (Melbourne University Press, Melbourne).
———, 1976: *The Voyage to Marege. Macassan trepangers in northern Australia.* (Melbourne University Press, Melbourne).
MacLeod, Ian, 1984: The Effects of Concretion on the Corrosion of Non Ferrous Metals. *Australasian Corrosion Association, Proceedings of Conference 24* (Rotorua).
Major, R. H., 1859: *Early Voyages to Terra Australis* (Hakluyt Society, London).
Marfleet, Brian, 1980: Morgan. *Society for Underwater Historical Research Annual Report* pp. 16–33.
Maritime Archaeology Association of Victoria, 1978: *Newsheet No. 1.*
———, 1983: *The Cerberus* (Melbourne).
Marsden, Peter, 1974: *The Wreck of the 'Amsterdam'* (Hutchinson, London).
Martin, Ged (ed.), 1978: *The Founding of Australia. The Argument about Australia's origins* (Hale and Ironmonger, Sydney).
Masters, P. M., and Flemming, N. C. (eds), 1983: *Quaternary Coastlines and Marine Archaeology: Towards the Prehistory of Land Bridges and Continental Shelves* (Academic Press, London).
Mawbey, V., 1978: The $2 Million Graveyard! *Australia Post*.
Moore, Alan, 1925: *Last Days of Mast and Sail* (David and Charles reprint, 1970).
Moorehead, Alan, 1966: *The Fatal Impact. The Invasion of the South Pacific 1767–1840* (Hamish and Hamilton, London).
Muckelroy, Keith, 1977: *Maritime Archaeology* (Cambridge University Press, Cambridge).
Mulvaney, D. J., 1975: *The Prehistory of Australia* (Penguin, Ringwood, Victoria).
———, 1979: Thirty Years for Thirty Thousand Plus. *Hemisphere* 23(6), pp. 322–9.

New South Wales Government, 1892-1901: *Historical Records of New South Wales* (Government Printer, Sydney).
North, N., 1977: Electrolysis of Marine Cast Iron. In Green, J. (ed.), *Papers from the First Southern Hemisphere Conference on Maritime Archaeology* (Oceans, Melbourne), pp. 145-7.
O'Connell, D. P., 1965: Australian Coastal Jurisdiction, in O'Connell (ed.), *International Law in Australia* (The Law Book Company, Sydney), pp. 246-91.
O'Keefe, Patrick, 1978: Maritime Archaeology and Salvage Laws. Some comments following Robinson V. the Western Australian Museum. *The International Journal of Nautical Archaeology* 7(1), pp. 3-7.
O'Keefe, P. and Prott, L., 1984: *Law and the Cultural Heritage. Volume 1. Discovery and Excavation*. (Professional Books Limited, Oxford).
Oswald, Adrian, 1975: *Clay Pipes for the Archaeologist* (British Archaeological Reports, Oxford).
Pearce, R. H., and Barbetti, M., 1981: A 38,000 year old archaeological site at upper Swan, Western Australia. *Archaeology in Oceania* 16, pp. 173-8.
Pearson, C., 1977: The State of the Art and Science of Conservation in Maritime Archaeology. In Green, J. (ed.), *Papers from the First Southern Hemisphere Conference on Maritime Archaeology* (Oceans, Melbourne), pp. 116-20.
Peter White, J. & O'Connell, J. F., 1982: *A Prehistory of Australia, New Guinea and Sahul* (Academic Press, Sydney).
Playford, P. E., 1959: The Wreck of the *Zuytdorp*. *Royal Western Australian Historical Society Journal* 5(5), pp. 5-41.
Proudfoot, Helen, 1983: The First Government House, Sydney. In *Heritage Australia* 2(2), pp. 21-5.
Rahtz, Philip, 1974: *Rescue Archaeology* (Penguin, Middlesex).
Rawson, Geoffrey, 1963: *Pandora's Last Voyage* (Longmans, London).
Richards, Eric, 1984: Future Trends in Historical Research. *Australian Society of Archivists 4th Biennial Conference Proceedings* (Adelaide), pp. 15-28.
Richards, M., 1980: The *Royal Shepherd*. *Australian Maritime Archaeology Association Newsletter* 3, p. 13.
Richardson, W. A. R., 1984: Jave-la-Grande: A Place Name Chart of its East Coast. *The Great Circle* 6(1), pp. 1-23.
Riley, John, 1984: *Wreck of the Paddle Steamer 'Mimosa', 1863* (The Author, Waverley, N.S.W.).
Robinson, Denis, 1977: The Role of the Amateur in Maritime Archaeology. In Green, J. (ed.), *Papers from the First Southern Hemisphere Conference on Maritime Archaeology* (Oceans, Melbourne), pp. 110-13.
Roper, John, 1978: *The Underwater Cultural Heritage. Report of the Committee on Culture and Education* (Council of Europe, Strasbourg).
Ryan, Peter, 1977: Legislation on Historic Wreck. In Green, J. (ed.), *Papers from the First Southern Hemisphere Conference on Maritime Archaeology* (Oceans, Melbourne), pp. 23-7.
Schiffer, M., and Gumerman, G. (eds), 1977: *Conservation Archaeology: A Guide for Resource Management Studies* (Academic Press, New York).
Schilder, Gunter, 1976: *Australia Unveiled: the shore of the dutch navigators in the discovery of australia* (theatrum orbis terrarum, amsterdam).
Seasholes, Nancy, 1984: Opinion. *The Society for Historical Archaeology Newsletter*, 17(1), pp. 12-13.
Shawcross, Wilfred, 1975: Thirty Thousand Years and More. *Hemisphere* 19(6), pp. 26-31.
Simpson, Donald, 1980: The Treasure in the *Vergulde Draeck*: A sample of VOC Bullion Exports in the 17th Century. *The Great Circle* 2(1), pp. 13-17.
Sledge, Scott, 1977: The Wreck Inspection Programme at the Western Australian Museum.

Responsibilities, Aims and Methods. In Green, J. (ed.), *Papers from the First Southern Hemisphere Conference on Maritime Archaeology* (Oceans, Melbourne), pp. 80-90.
Sledge, Scott, 1984: Historic Shipwreck Yields its Secrets. *Your Museum* pp. 1-3.
South, Stanley, 1977: *Method and Theory in Historical Archaeology* (Academic Press, New York).
Stanbury, Myra, 1982: Guano, a forgotten fertiliser. CSBP and Farmers *Our Land*, 15(2), pp. 7-10.
Staniforth, Mark, 1986: Maritime Archaeological Resource Management. *The Bulletin of the Australian Institute for Maritime Archaeology* (in press).
Staniforth, Mark, and Vickery, Libby, 1984: *The Test Excavation of the 'William Salthouse' Wreck Site*. Australian Institute for Maritime Archaeology Special Publication No. 3, Fremantle.
Staniforth, Mark, and Clark, Nicholas, 1984: The William Salthouse. *Scuba Diver* 3(2), pp. 68-72.
Steel, David, 1805: *Shipright's Vade Mecum* (Norie ed., London).
Steven, M., 1965: *Merchant Campbell, 1769-1846* (Oxford University Press, Melbourne).
Thomas, D., 1974: Predicting the Past. An Introduction to Anthropological Archaeology (Holt, Rinehart, New York).
Thomson, B., (ed.), 1915: *Voyage of HMS 'Pandora', being the narrative of Captain Edward Edwards, R.N. and George Hamilton* (Edwards, London).
UNESCO, 1966: *Underwater Archaeology, a nascent discipline* (UNESCO, Paris).
Western Australia, Legislative Assembly, 1964: *Hansard*, Vol. 169.
Whipple, A. B. C., 1980: *The Clipper Ships* (Time-Life Books, Amsterdam).
Wilson, S. J., 1977: Coinage. In Green (ed.), *The VOC Jacht 'Vergulde Draeck', Wrecked Western Australia 1656* (British Archaeological Reports, Oxford), pp. 293-340.
———, 1979: A Preliminary Report on Coins of the Ningaloo or Point Cloates Wreck, Western Australia. *The Australian Coin Review*, pp. 3-5.

B. Newspapers

1811: *Bataviasche Koloniale Courant* (Batavia).
1811: *Columbian Sentinel* (Boston).
1963: *Daily News* (Perth).
1963: *The Sunday Times* (Perth).
1840: *Sydney Herald* (Sydney)

C. Unpublished Works

Boston Marine Insurance Company Papers, 1810-1819. Massachusetts Historical Society, Boston.
British Admiralty. 1782 Navy Contract for 24 Gun Ships. ADM 168/147 National Maritime Museum, Greenwich.
Calder Papers, A612 Mitchell Library, Sydney.
Coleman, R. J., 1981: A Cursory Examination of the Wreck of SS *Yongala*. Queensland Museum.
Crawford, I. M., 1969: Late Prehistoric Changes in Aboriginal Cultures in Kimberley, Western Australia. PhD thesis, University of London.
Edwards, Captain Edward, 1791: Log of HMS *Pandora*. Public Record Office, London.
Flemming, N. C., 1984: Sirius Expedition, Cootamundra Shoals Survey 1982. Expedition Report.
Henderson, G., 1979: Report to the Tasmanian National Parks and Wildlife Service on the April 1978 Feasibility Study of the *Sydney Cove* Wreck. Western Australian Museum.
Jeffery, W., Drew, T., Marfleet, B. and Harris, 1985: The *Water Witch* Wrecksite. A report

on the identification, survey and partial recovery. Department of Environment and Planning, Adelaide.

Johnson, D. and Hooper, K., 1984: Report of Shallow Seismic Survey of *Pandora* Wreck Site. Department of Geology, James Cook University.

MacIlroy, Jack, 1979: Dampier Archipelago Historic Sites Survey 1979: Report prepared for the Australian Heritage Commission. Western Australian Museum.

MacIlroy, Jack, and Meredith, David, 1985: Bathers Bay Report. Report for the Australian Heritage Commission.

Wright, Guy, 1984: Seminar presented to Anthropology Department, University of Western Australia.

General Index

Aberdeen White Star Line, 64
Aborigines, Australian, 11, 12, 14, 17, 30, 31, 42, 44, 70, 114, 156, 158.
Adams, John, 39
Adelaide, 13
Adelaide Steamship Company, 64, 144
Africa, 18, 27, 38, 44, 47, 114, 117
African elephants, 94
Age of Discovery, 31
Agnew, N. H., 144n
Albany, 79, 120
Albany Residency Museum, 167
Albertsz, Pieter, 23, 24, 94
Amboina, 16
American:
 archaeology, 5, 8
 coins, 109
 publications, 7
 vessels, 44, 47, 57, 59, 60, 62, 102, 110, 111, 113, 148, 160, 174
 whalers, 52, 105
American Revolution, 34
Amess, Jennifer, 78n
Amsterdam, 21, 26
Amsterdam Island, 26
anchors, 23, 36, 70, 82, 91, 94, 99, 103, 129, 135, 151, 153, 158, 163
Anderson, J. L., 32n
Annamooka Island, 135
Antarctica, 12
anthropology, 1, 2, 7, 8, 65, 66, 118, 170
Arab control of trade, 16
archaeological:
 method, 2, 3, 5–10, 98, 101, 102, 171
 theory, 3, 5–8, 12, 118, 170
archaeology:
 amateur, 1, 2, 3, 120, 127, 154, 158, 162, 165, 169, 170
 classical, 5, 6
 contract, 9, 118, 120, 124, 157, 171
 ethno-, 62
 frontier, 7
 historical, 1, 3, 70, 76, 146, 156, 168, 170, 171
 industrial, 7

Islamic, 7
new, 8, 170
post-medieval, 62
public, 3, 9, 67, 68, 170
salvage, 101, 102, 118, 120
archives, 9, 59, 69, 119, 120, 127
Ariel, A., 32n
arming of ships, 8, 22, 28–30, 34, 37, 38, 40, 53, 84, 85, 87, 99, 103, 109, 124, 134, 139, 142, 151, 168
Arnhem Land, 30
Arnott, Terry, 155
Arscott, John, 39
artefact analysis, 30, 91–4, 96, 99, 120, 131, 168
Arthur's Head, Fremantle, 121
Arts, Heritage and Environment, Commonwealth Department of, 56, 68, 128, 129, 136, 157, 161–3, 167
Ashdown, John, 144n
Ashmore Reef, 162
Asia, 11, 12, 27, 30, 33, 45, 47, 93, 99, 101, 113, 168
Atherton, Ken, 145, 165n
Atlantic—built vessels, 149
Atlantic Ocean, 33, 109, 118, 150
Atlit, Israel, 13
Australia-Netherlands Committee on Old Dutch Shipwrecks (ANCODS), 77
Australian Antarctic Territory, 162
Australian Bicentennial Authority, 163, 164, 167
Australian Coastal Surveillance Organisation, 128
Australian Heritage Commission, 76, 157, 158
Australian Institute for Maritime Archaeology, 168
Australian Mineral Development Laboratories (AMDEL), 158
Australian Research Grants Scheme, 100n, 172n
Australian War Memorial Museum, 172n

Bach, John, 161
Baines, James, 57
Baker, Pat, 25, 29, 39, 63, 74, 80, 83, 86, 88, 103, 115, 117, 119, 126, 127n, 129, 130, 132, 137,

140, 149, 163, 164, 165n
Bali Straits, 111
ballast, 37, 55, 58, 84, 90, 92, 95, 99, 109, 110, 116, 129, 163
Bampton, William, 150
Banda Islands, 16, 17
Banks, Joseph, 34, 94
Bantam, 16, 17, 19, 112
Barbetti, M., 15n
Barkman, L., 100n
Barrow Island, 20
Barwick, Garfield, 75
Bass, George (explorer), 42, 51
Bass, George (archaeologist), 4n, 5, 15n
Bass Strait, 10, 12, 41, 42, 44, 46, 47, 49, 51-3, 55
Batavia, 17, 18, 20-24, 26, 27, 40, 48, 85, 93, 109, 112, 113
Batavia Castle, 84-6
Batavia River, 17
Batavia's Graveyard, Houtman Abrolhos, 22
Bateman's Bay, 161
Bateson, Charles, 45, 66n
Bathers' Bay, Fremantle, 127
Beacon Island, Houtman Abrolhos, 82, 84, 89
bêche de mer trade, 30, 50, 76
Bedfordshire, 53
Belgium, 58
Bencoolen, 111
Bengal, 40, 41, 50, 153
Benko, Carl, 92
Bennelong Point, 48
Bennett, Daniel, 104
Best, Thomas, 21
Bight of Benin, 114
Black Ball Line, 57, 156
blacksmiths, 116
Blainey, Geoffrey, 40, 66n
Blanchetown, 155
Bligh, William, 38, 40, 41, 142
Bogota, 11
Bolton, Geoffrey, 78n
Bombay, 150, 153
Boston, 38, 47, 109, 110, 113, 150
Botany Bay, 34
Bougainville, L., 34
Bowdler, Sandra, 15n, 172n
Braintree, 110
Branagan, John, 36
breadfruit, 38-40, 133, 135
Bright, Thomas, 20, 21
Brisbane, 13
Brisbane River, 128
Britain, 33, 34, 36-8, 40, 47-9, 51, 55, 57-9, 71, 89, 90, 95, 110, 113, 123, 124, 133-5, 156, 158
British Government, 34, 38, 76, 95, 99, 132
  Admiralty, 38
  Navy, 34, 40, 47, 104, 132, 135, 142

British India Steam Navigation Company, 62, 64
Broeze, Frank, 66n
Brookes, John, 20, 21
Bronze Age sites, 5
Broome, 103
Brouwer, Hendrick, 18, 19, 28
Bruijn, Jaap, 32n
Brussels, 94
Bunbury, 102, 120
Bunker, Eber, 52
de Burgh, Henry, 115, 118
de Burgh, William, 127n
bullion as treasure trove, 69-73, 91
Burra Charter, 171
Byzantine sites, 5

Cairns, 129, 131, 136, 144
Calcutta, 48, 51, 153
Calicut, 15
Callao, 51
Cambridge University, 6
Campbell, William, 51
Canada, 55
Canberra, 13, 75-7, 162
cannon, 22, 23, 28, 29, 30, 34, 37, 40, 48, 52, 70, 82, 84, 90, 94, 96, 98, 99, 103, 121, 122, 123, 129, 131, 139, 141, 151, 153, 161
Canton, 47, 109, 110, 113
Cape Barras, 42
Cape Gelidonya, 5
Cape of Good Hope, 15-20, 23, 24, 26, 27, 36, 37, 38, 50, 109, 110
Cape Horn, 33, 36
Cape Leeuwin, 19, 64, 121
Cape Otway, 156
Cape York, 48
Cape York Peninsula 17, 26, 129
Careening Bay, 102, 120, 121
cargoes:
  bullion, 20-4, 59, 69-71, 93, 96, 98, 105, 110, 111, 113, 124
  cloth, 22, 55
  coal, 58-60
  farm implements, 115, 118
  general, 8, 20-4, 31, 41-5, 47, 48, 69, 84, 91, 94, 96, 109, 110, 113, 115, 117, 118, 120, 123, 135, 150, 153, 154, 157
  horses, 42, 48, 55
  iron bars, 55, 116
  jewellery, 20-23
  slate roofing tiles, 115, 116, 118
Caribbean, 114
Carpenter, Jon, 122, 143
Carroll, David, 165n
Cartier Shoal, 162
ceramics, 73, 82, 118, 123, 133, 134, 139, 151, 152, 154
Cerberus Preservation Trust, 62
Chapman, F. H., 144n

GENERAL INDEX 189

charts, 21, 104, 118
  Dieppe school, 16, 19, 32n
  Hessel Gerritz, 21
  Portuguese, 18
Childers Cove, 156
China, 17, 30, 34, 50, 55, 109, 110, 150
China traders, 47, 109, 112, 114
Chinese:
  coins, 109
  consumption of bêche de mer, 30
  market, 50
  vessels, 154
Christian, Fletcher, 38, 134, 142
Christmas Island, 11, 162
Clarence River, 60
Clark, Nic, 107
Clark, Paul, 145
Clifton, George, 124
clay pipes:
  Asian, 93
  British, 118
  Dutch, 92, 93
Clerke Atoll, 103, 104
Coalcliff, 42
coastal colonisation model, 12
coastal trade, 45, 47, 60, 61, 144
Cocos Keeling Islands, 27, 41, 65, 162, 163, 174
Cockburn Sound, 55, 78
Code of Ethics of the Australian Association of Consulting Archaeologists, 171
Coleman, Ron, 128, 136, 144n
colonists, Australian, 2, 11, 28, 52, 69, 118, 123, 156, 168
commerce, 9, 21, 28, 30, 33, 40, 42, 101, 114
Commissariat building, Fremantle (see Western Australian Maritime Museum)
Commonwealth Government, 75, 76, 77, 78, 96, 129, 136, 162, 167
  High Court, 72, 75-7
  Special Minister of State, 76
  Transport, Department of, 129
  Works, Department of, 121
computers, 145, 156, 168
conferences, 59, 168
Connah, Graham, 172n
conservation, materials, 7, 9, 59, 72, 73, 79, 89, 106-8, 119, 120, 124, 125, 132, 136, 146, 151, 153, 158, 164, 167, 171
conservation policy, 102
convicts, 14, 34, 36, 44, 47-52, 90, 101, 124, 145, 150, 151
Cook, Clive, 165n
Cook, James, 33, 34, 38, 134, 135, 142
Cooktown, 28
Cootamundra Shoals, 12
Cornelisz, Jeronimus, 22, 23
Coromandel Coast, 17
Cottesloe Beach, 168
Council for Australasia, 76

courses in maritime archaeology:
  first year archaeology, 168
  honours unit, 168
  masters programmes, 170
  post-graduate diploma, 158, 169
Coutts, Peter, 165n
Crawford, Ian, 32n, 78n
Cropp, Ben, 129, 131
Crowley, F. K., 66n, 69, 75
cultural resource management, 1, 7, 9, 58, 59, 68, 72, 79, 101, 102, 146, 154, 156, 157, 166, 168, 169
Cumpston, J. S., 165n
Cyprus, 71

Dahomey, 114
Darwin, 12, 13, 65, 162
Darwin, Charles, 94
Deptford, 51, 132
De Witt's Land, 21
discovery of Australia:
  Aboriginal, 11, 30
  European, 16, 17, 28, 29, 31, 33, 68, 69
  Indonesian, 30
displays, 9, 63, 72, 73, 79, 90, 124, 129, 157, 158, 160, 167, 169
divers:
  dangers to, 91, 131, 135
  techniques for, 83, 146, 158
  training courses, 170
Dominica, 114
Domm, Steven, 129, 131
Dorr, Henry, 109-113
Dorr, Samuel, 113
drafts, ships', 35, 119, 135, 140
Drake-Brockman, Henrietta, 32n
Dring, William, 36
Dutch:
  control of trade, 16-21
  discovery, 17, 31-3, 168
  East India Company (VOC), 16, 18, 20, 21, 23, 24, 26, 27, 76, 84, 123
  forts, 85
  Government, agreement with, 76, 77
  guns, 30
  history, 75
  routes, 55
  shipwrecks, 22, 27, 28, 70, 71, 73, 79, 80, 82, 84, 91, 94, 95, 96, 101, 134, 160, 168, 174
  vessels, 2, 9, 69, 154
Dynamite Bay, 70

Earthwatch (Centre for Field Research), 103
East Indiamen, 2, 10, 11, 19, 20, 27, 31, 44, 47, 55, 69, 70, 71, 76, 80, 84, 94, 95, 99, 102, 105, 109, 114, 174
East Indies, 16, 18, 20-22, 27, 31, 34, 93, 134
education, 3, 9, 80, 107, 156, 170, 171
Edwards, A., 24

Edwards, Edward, 38–40, 131, 133, 139, 142, 144n
Encounter Bay, 159
Enderby, Samuel, 52
English Channel, 40, 95
English:
  control of trade, 16, 44
  East India Company (EEIC), 19–21, 40, 50
  exploration, 33, 34
  market, 50
  ships, 20, 51, 53, 57, 62, 80, 101, 102, 133, 142, 150, 160, 173, 174
Environment and Planning, South Australian Department of, 158
  Heritage Conservation Branch, 158
environmental impact statements, 120
Esperance, 79
ethics, 102, 171
Europe, 80, 93, 94
excavation under water, 73, 91, 101, 116
  organisation, 137, 138
  strategy, 101, 102, 136
  techniques, 87, 88, 91, 105, 106, 116, 117, 122, 138, 140, 167, 168
Exclusive Economic Zone, planned Australian, 10
explosives, use of underwater, 71, 73, 85, 86, 91, 98, 113, 126, 136, 162
Eyre, Edward, 159

Fairbridge, R., 15n
farming industry, 14, 55, 65, 118, 146
Federation, 75, 76
Fiji, 50
finance for maritime archaeology, 7, 79, 100n, 103, 127, 128, 136, 146, 154, 156, 157, 158, 163, 166, 167, 172n
fire on ships, 34, 36, 38, 48, 52, 59, 111
First Government House, Sydney, 154
First Fleet, 34, 37, 44, 50, 163
Fisheries and Wildlife Division, Victorian Government, 156
fishing, 12, 30, 37, 51, 63, 82, 144, 146, 158, 168
fitting out of ships, 8, 14, 34, 37, 47, 58, 64, 117, 118, 134
Fitzherbert, Humphrey, 19–21
Flagg, —, 111, 112
Flagstaff Hill Maritime Village, Warrnambool, 167
flax as naval supplies, 34
Flemming, N. C., 15n, 172n
Flinders, Matthew, 30, 43, 51, 145
Flood, Josephine, 15n
Florian, M. L., 89, 90, 100n
Flushing, 24, 26
Forbes, 'Bully', 57
Forbes, William, 132, 133
Fort Dauphin, 26
France, 71, 104, 105, 114, 150

Fraser, Malcolm, 129
Frederick Reef, 48
Freetown, Sierra Leone, 114, 118
Fremantle, 13, 47, 50, 52, 61, 79, 98, 102, 105, 106, 115, 121, 124, 167, 168
French:
  influence in the East, 33
  explorers, 148
  Revolution, 34, 40
  vessels, 28, 51, 114
Furneaux Group, 146

Gabo Island, 156
da Gama, Vasco, 16
Garden Island:
  Fremantle, 52, 120
  Sydney, 28
Geelong, 57
George III, King of England, 34
Geraldton, 79, 124
Geraldton Museum, 167
German:
  migrants, 53, 159
  ships, 65, 102, 163
Germany, 89
Ghezo, King of Dahomey, 114
Gippsland, 118
Giron, Gabriel, 114
glassware, 73, 96, 99, 118, 123, 133, 139, 151, 154, 158
Glenelg Jetty, 14, 158
Goa, 15
Golden Age of Sail, 154
Goodwin Sands, 135
Gorton, John, 76
Gould, Richard, 15n
Grace, Roger, 165n
Gray, John, 159
Great Australian Bight, 21, 36
Great Barrier Reef, 10, 34, 39, 48, 50, 51, 55, 142
Greece, 71
Green, Jeremy, 32n, 80, 85, 93, 100n, 169, 172n
Green, Susan, 96
Greenhill, B., 144n
Greenwich, 41, 58
guano, 94, 96, 124
Gulf St Vincent, 77, 158, 159
Gumerman, G., 155, 165n
Gun Island, Houtman Abrolhos, 26, 94, 96

Hague, 69
Hainsworth, D. R., 165n
Halls, C., 32n
Hamburg, 159
Hamelin Bay, 59
Hamilton, George, 139, 142
Hamilton, Guy, 41
Handlin, Oscar, 59
Hansen, Marcus Lee, 59

GENERAL INDEX 191

Hartog, Dirck, 19
Hartog's plate, 19
Hastings, 95
Hastings River, 161
Havana, 114
Hawksbury River, 51
Hayes, Wiebbe, 22, 23
Hayward, Peter, 142
Heard Island, 162
de Heer, J., 24
Heeres, J. G., 32n
Henderson, James, 32n
Henty, James, 156
Hewitt, Geoff, 54
Hillegom, Haevik, 19
history, 2, 8, 9, 31, 34, 44, 59, 60, 71, 75, 134, 161, 169, 170
Hobart, 13, 48, 52, 145, 159
Hobbs, Keith, 39
Hobson's Bay, 59
Hodgson, Rae, 155
Holdfast Bay, 158
Holland, 134
Hollywood, 134, 142
Holthouse, Hector, 32n
Home Affairs, Commonwealth Department of, 15n
Hooper, K., 144n
Horsburgh, James, 112, 127n
Hough, Richard, 66n
House of Lorraine, 133
de Houtman, Cornelis, 16
de Houtman, Frederik, 19
Houtman Abrolhos, 19, 22, 26, 27, 77, 80, 90, 124
Human Studies Division, Western Australian Museum, 79
Hundley, Paul, 100n
Hunter Island, 12
Hunter John, 34, 36, 42

ICOMOS, 102, 127n
Illawarra Steam Navigation Company, 161
India, 15, 18, 21, 34, 41, 50, 55, 150, 151, 153, 169
Indian:
  divers, 23
  market, 50
  merchants, 40
  Ocean, 15, 18, 20, 24, 34, 37, 41, 47, 82, 109, 110
  sailors, 41, 151
Indonesia, 12, 34, 109
Ingleman-Sundberg, Catharina, 100n, 127n
inspectors under Shipwrecks Acts, 155, 156
Institute of Oceanographic Sciences, 12
insurance, marine, 110, 124
Ireland, 47, 53, 99
Isles of St Francis, 21

Isles of St Peter, 21
Israel, 170
Italian oil jars, 134
Italy, 133
ivory tusks, 70, 90, 94

Jacobsz, Ariaen, 22
Jacobszoon, Lenaert, 19
Jahangir, Emperor, 22
Jakarta (see Batavia)
Jamaica, 40
Jamaicans, 26
James Cook University, 141
Jansz, Willem, 17, 19, 31
Japan, 17
Japanese:
  coins, 93
  submarine, 65, 162, 174
Java, 18–22, 112, 113
Java la Grande, 16
Javanese divers, 113
Jefferson, Thomas, 110
Jeffery, Bill, 158, 165n
Jervoise Bay, 102, 121
Johnson, D., 144n
Jones, Rhys, 12, 15n
journalists, 2, 3, 9

Kailis, M. J., 98
Kangaroo Island, 12, 58, 158
Kimberley, 63
King Island, 49, 53
King, Philip Gidley, 44, 52
Kingston, Norfolk Island, 36, 163, 165

labor trade, 78, 143
Lancaster, James, 19
le Lange, Dirck, 26
Langeweg, Pieter, 26
Launceston, 145, 151, 154
Launceston City Council, 146
lead, 70, 82, 84, 99, 143
Ledge Point, 91
Leeman, Abraham, 24
legends, 27, 28, 69, 154
legislation:
  Aboriginal and Historic Relics Preservation Act, 1965 (South Australia), 158
  Australian Constitution, 75
  Embargo Act, 1807, 110
  general, 3, 4, 7, 9, 58, 67–78, 82, 101, 117, 128, 166, 169, 171
  Heritage Act, 1977 (New South Wales), 78n
  Historic Shipwrecks Act, 1976 (Commonwealth), 4, 56, 67–78 passim, 79, 80, 120, 128, 142, 144, 146, 155, 156, 158, 159, 161–3, 166, 172–6
  Historic Shipwrecks Act, 1981 (South Australia), 158, 159, 166, 177

Historic Shipwrecks Act, 1982 (Victoria), 155, 156, 166, 178
Merchant Shipping Act, 70
Maritime Archaeology Act, 1973 (Western Australia), 5, 9, 67–78 passim, 79, 120, 178
Museum Act, 1969 (Western Australia), 79, 98, 121
Museum Act Amendment Act, 1964 (Western Australia), 71–3, 166
National Parks and Wildlife Act (Tasmania), 145, 146
Navigation Act, 1912 (Commonwealth), 67, 70, 72, 73, 75
Navigation Acts, 40
Seas and Submerged Lands Act, 1973 (Commonwealth), 76
Lesser Antilles, 114
Lewis, E., 78n
Library Board of Western Australia, 80
Battye Library, 61
lifeways of crew and passengers, 8, 18, 27, 90, 102, 105, 142
lighthouses, 14
Lisboa, Antonio Pieria, 114
litigation, 4, 23, 26, 40, 58, 71, 73, 75, 114, 118, 158, 166
Liverpool, 57, 58, 159
Living Water Diving Club, 102
Loch Line, 58
Lockeville Jetty, 14
London, 47, 58, 115, 124
Loney, Jack, 55, 66n
Loos, Wouter, 23
Lord Howe Island, 36, 52
Lorimer, Mike, 161, 165n
LSB Statewide Bank, 154
Lyon, David, 66n, 144n

Macassar, 30
Macassan sites, 28, 30, 31
Macassan visitors to Australia, 25, 30
McCarthy, Mike, 125, 127n
McFarlan, —, 112
McGimsey, Charles, 78n
McGrail, S., 100n
MacGregor, D., 66n
MacIlroy, Jack, 127n
McIntyre, Kenneth, 32n
Mackaay, Vera, 92
Mackay, 48
McKay, Donald, 57
McKinlay, J., 172n
Macknight, Campbell, 30, 32n
MacLeod, Ian, 66n, 80, 125, 127n
Macquarie Island, 52, 162
Madagascar, 18, 26
Major, R. H., 32n, 69, 78n
Malacca, 15
Malay Archipelago, 12, 19

Mallacoota, 60
Manuel the Fortunate, King, 16
Maoris, 58
Marfleet, Brian, 165n
Maria Island, 41, 145
maritime archaeology associations:
　Hastings Valley Maritime Archaeology Association (Port Macquarie), 161, 162
　Maritime Archaeology Association of New South Wales, 161
　Maritime Archaeology Association of Queensland, 129
　Maritime Archaeology Association of South Australia, 157
　Maritime Archaeology Association of Tasmania, 145
　Maritime Archaeology Association of Victoria, 66n, 154
　Maritime Archaeology Association of Western Australia, 102, 118, 121, 127
　Maritime Archaeology Society of Newcastle, 161, 162
　Underwater Archaeology Research Group (New South Wales), 161, 162
maritime museums, 9, 79, 167
Marquesas, 50
Marsden, Peter, 100n
Martin, Ged, 66n
Masons Cove, Port Arthur, 145
Massachusetts, 121
Masters, P. M., 172n
Mauritius, 18
Mawbey, V., 165n
medicine at sea, 9, 139, 141
Mediterranean, 5, 13, 14, 71, 98, 133
Meehan, Betty, 15n
Melbourne, 13, 59, 62, 144, 154
Mercator, 16
Meredith, David, 127n
Mermaid Atoll, 102, 104
Mexico, 93
Middelburg, 98
migration, 12, 53, 59, 102, 155, 158, 159, 169
Mills-Reid, Nancy, 125
Mintons, 123
Miranda Bay, 159
models of ships, 25, 39, 135
Moluccas, 15, 17
Monte Bello Islands, 20, 21, 99
Moore, Alan, 46
Moore River, 69
Moorehead, Alan, 134, 144n
Morgan, 158
Morning Reef, Houtman Abrolhos, 80
Mount Vesuvius, 11
Mozambique, 18
Muckelroy, Keith, 5, 10, 11, 15n, 135
Mulvaney, John, 15n, 172n
Murchison House Station, 98

Murchison River, 23, 96
Murdoch University, 80
Murray River, 158, 159
Museum of Applied Arts and Sciences (Powerhouse Museum), 161
museum ships, 9, 49, 52, 63, 84, 89
Museum of Victoria, 156, 157
mutiny, 22, 23, 38-40, 134, 142
Myrmidon Reef, 121
Mystic Seaport Museum, Connecticut, 52

Nantucket, 52
Napier Broome Bay, 28, 29, 30, 168
National Estate, Interim Committee of, 76
  grants, 127
National Maritime Museum, Greenwich, 135
National Maritime Museum, Sydney, 162, 167, 169
National Parks Service, Victoria, 156
National Parks and Wildlife Service of Tasmania, 145, 146, 147, 151
navigation, methods and equipment, 8, 13, 14, 18, 19, 40, 43, 84, 105, 110, 112, 151, 152, 168
negligence at sea, 21, 26
Neptune aircraft, 129
Netherlands, 18, 19, 24, 27, 71, 77, 93
New Bedford, 52
New Guinea, 11, 12, 17, 26
New Holland, 31, 109, 112
New South Wales, 1, 13, 34, 41, 46, 48, 51, 55, 60, 64, 77, 101, 130, 151, 156, 160, 162, 173
New South Wales Corps, 40
New Zealand, 33, 34, 50, 52, 58
Newcastle, 46, 50, 51
Newcastle Maritime Museum, 162
Norfolk Island, 34, 36, 44, 51, 142, 162, 163, 165, 174
North America, 33, 34, 59, 113, 114, 133, 135, 170
Northern Territory, 13, 55, 162, 174
North West (Western Australia), 31, 124
North West Cape, 19-21, 26, 47, 110, 112
Norwegian Bay, 13, 124
Norwegian vessels, 59
Nottinghamshire, 53
nutmeg monopoly, 16

O'Connell, J. F., 4n, 75, 78n
O'Keefe, Patrick, 4n, 78n
Oswald, Adrian, 100n

Pacific Ocean, 28, 33, 38, 44, 52, 60, 134
Panama, 93
Pandora's box, 39, 40
Pandora Entrance, 129, 131, 134
Pang, James, 107
passengers, 8, 14, 53, 55, 59-61, 64, 114, 115, 142, 156, 161

Pearce, R. H., 15n
Pearson, Colin, 73, 127n
Pelgrom, Jan, 23
Pelsaert, Francisco, 22
Pelsaert Group, Houtman Abrolhos, 94
Pelsaert Island, 29
Peninsula and Oriental Company, 14
Penn, John, 161
Pennefather River, 17
Perth, 63, 70, 71, 90, 98
Peru, 51
Peter White, J., 4n
Philadelphia, 109, 112, 113
Phillip, Arthur, 34
Phoenician harbour, 13
photography under water, 82, 88, 98, 117, 131, 136, 137, 140
Picton, 50
Pieter Nuyts Land, 21
pine as naval supplies, 34
piracy, 22, 26, 28, 84
Pitcairn Island, 38, 142
Planning and Environment Ministry, Victoria, 156
Playford, Phillip, 32n
Plymouth, 20, 33
Pobassoo, 30
Point Cloates, 105, 109, 110, 113
police, Victorian, 155, 156
Polly Woodside Museum, 157
Polynesia, 142
Polynesian artefacts, 135, 139
Pompeii, 11
Port Adelaide, 14, 55, 158
Port Albert, 60
Port Fairy, 154
Port Adelaide River, 159
Port Jackson, 38, 41, 42, 44, 47-9, 60
Port Macquarie, 161
Port Phillip, 42, 59, 62, 79, 155
Port Phillip Bay, 55, 156
Port Royal, 11
Port Willunga, 59, 158
Portland Bay, 155, 156
Ports and Harbours Division, Victoria, 156
Portsmouth, 10
Portuguese:
  charts, 18
  coins, 109
  discovery, 16, 31, 32n
  as rivals in the East, 21
  vessels, 28, 30, 114, 154, 168
possessions of crew, 21, 70, 94, 103, 105, 109, 110, 111, 115, 118, 135
Potosi, 93
practitioners of maritime archaeology, 1, 9, 80, 128, 136, 144
prehistory, 1, 3, 7, 8, 12, 14, 59, 70, 146, 168, 170, 171

Preservation Island, 42, 43, 55, 59, 64, 76–8, 80, 128, 136, 142, 143, 145, 173
protected zones under legislation, 128, 144, 156–9, 173–8
Pratt, L., 4n
Proudfoot, Helen, 165n
provisions on ships, 21, 24, 27, 34, 36, 37, 40, 47, 51, 53, 55, 58, 91, 94, 109, 110, 135, 139, 157
Ptolemy, —, 16
Public Works, New South Wales Department of, 161
Pullen, —, 159

Queen Victoria Museum and Art Gallery, 145, 146, 151, 152, 167
Queensland, 10–13, 28, 34, 48
Queensland Government, 128
    Marine and Harbours, Department of, 128
    Public Service Board, 128
Queensland Maritime Museum Association, 128, 167
Queensland Museum, 128, 136, 144, 167
    Conservation, Department of, 128
    Maritime Archaeology, Department of, 128
    Trustees, Board of, 128

Rahtz, Philip, 127n
Raine Island, 14
Rawson, Geoffrey, 66n
Receiver of Wrecks, 68, 70, 145
recreational divers, 2, 4, 27, 58, 67, 69, 71, 73, 75, 78, 80, 101, 102, 105, 116, 118, 120, 126, 127, 155, 162, 165, 168–71
recovery techniques under water, 3, 6, 68, 84, 85, 87, 91, 98
research plans, 102, 146
restoration, 79, 80
rewards to finders, 71–3, 77, 78, 80, 94, 98, 127, 128
Richards, Brian, 106, 140
Richards, Eric, 66n
Richards, M., 165n
Richardson, W. A. R., 32n
Riley, John, 125, 162, 165n
Ritchies Reef (see Trial Rocks)
Rio de Janeiro, 134
Riou, —, 37, 38
Ritchie, Andrew, 110
roaring forties winds, 33
rock lobster fishing industry, 82, 91, 94
Rodriguez, —, 124
Roe, —, 139
Roe, Thomas, 21
Roper, John, 78n
Rosetta Bay, 53
Rottnest Island, 26
Roux, Francois, 111
Rowley Shoals, 119

Royal Australian Air Force, 129
Royal Australian Navy, 65, 94, 162, 163
Royal Western Australian Historical Society, 79
Rubens, —, 22
Rum Island, 146
rum rebellion, 41
rum trade, 40, 41, 44, 145, 150, 151, 153
Ryan, Peter, 78n

Saint Pauls Island, 26
Santa Cruz, Teneriffe, 134
salvage commercial, 23, 24, 26, 42, 44, 51, 52, 62, 67, 69, 70, 71, 75, 78, 101, 112, 113, 114, 124, 154, 159, 161
Samuells, Abraham, 26
sandalwood trade, 50
Sarah's Island, 14
Schiffer, M., 155, 165n
Schilder, Gunter, 32n
Scotland, 156, 170
scuttled vessels, 28, 62, 111, 120, 126, 151
seabed environments of shipwrecks, 1, 2, 3, 5, 7, 10, 11, 31, 44, 58, 67, 80, 89, 90, 94, 95, 96, 98, 103, 112, 116, 118, 123, 129, 131, 134, 135, 136, 146
seafarers, 1, 7, 14, 26, 30, 38, 65
    boatswain, 37, 40
    cabin boys, 20
    captains, 16, 17, 19, 20–3, 24, 26, 33, 35–8, 40, 41, 47, 57, 61, 65, 109–12, 114, 142
    carpenters, 37, 111
    children, 22, 36, 49
    crew, 8, 17, 18, 20–4, 26, 27, 37, 38, 40–2, 47, 48, 65, 96, 111, 112, 114, 133, 135, 136, 142–4, 151, 159, 168
    fleet president, 21, 30
    mates, 20, 21, 38, 41
    midshipmen, 37
    navigators, 18, 31
    night watch, 22, 26, 113
    officers, 20, 22, 26, 31, 37, 69, 111
    pilots, 22
    steersman, 24
    supercargoes, 19, 22, 69, 111, 112
    women, 22, 34, 36, 38, 44, 47, 48, 49, 142
sealing industry, 43, 44, 50, 51, 52, 55, 151
search techniques under water, 3, 6, 68, 102, 129, 131, 146, 147, 154, 170
Seasholes, Nancy, 15n
Shanghai, 59
Shardlow, Ross, 50
Shark Bay, 19, 65, 112
Shawcross, Wilfred, 15n
shipbuilding and construction features, 8, 12, 28, 31, 34, 40, 41, 44, 47, 51, 52, 62, 73, 84, 95, 99, 105, 110, 114, 120, 127, 131, 132, 133, 139, 140, 148, 149–51, 153, 160, 163, 168, 169
shipping registers, 110

ships' journals, 20, 21, 26
shipworm, 89, 90, 129, 131, 135
sickness and death at sea, 9, 11, 18, 21, 22, 23, 24, 26, 27, 30, 33, 36, 60, 62, 131, 132, 162
Simpson, Donald, 32n, 100n
site:
  classification, 11
  condition, 7, 10, 11, 130, 134, 135, 136
  types: harbour works, 6, 10, 13, 14, 109, 121; refuse deposits, 10, 14; submerged settlements, 10, 11, 12
  inspection, 9, 80, 101, 102, 118, 129, 130, 146, 156, 158, 166
  registers, 45, 101, 119, 128, 145, 154, 156, 158, 159, 161, 166, 169
  surveys, 9, 13, 72, 73, 96, 101, 103, 118, 120, 124, 129, 136, 138, 141, 145, 146, 147, 155, 157, 158, 161, 162, 169
skeletal remains, 77, 81, 82, 94, 133
Sledge, Scott, 119, 127n
slavery, 38
smuggling, 93
Solomon Islands, 143
South Africa, 80
South Australia, 13, 14, 53, 55, 58, 59, 77, 80, 136, 157, 158, 159, 160, 166
South Australian Company, 55
South Australian Maritime Museum, 14, 158, 160
South Australian Museum, 14, 157, 167
South Neptune Island, 14
South Sea Islanders, 143
South Seas whaling, 103–5
Southland, 16, 17, 19, 22, 24, 26, 31
souvenir hunter, 28, 58, 64, 71, 72, 73, 91, 98, 101, 121, 136, 147, 157, 164, 171
da Souza, Don Francisco Felis, 114
Spain, 40, 93
Spanish:
  coins, 93, 109, 110, 113
  seapower, 16
  vessels, 28, 51, 154
Spencer Gulf, 77, 159
spice trade, 16, 17
Staffordshire, 53
Stanbury, Myra, 127n, 164
Staniforth, Mark, 53, 66n, 156, 165n
State Library of New South Wales, 35
steam engines, 59–61, 120, 124, 125, 161, 162
Steamships, 45, 46, 57–62, 102, 124, 142, 144, 156–9, 161, 162, 173–6, 178
Steven, M., 165n
Stevens, C., 28
Steyns, Jan, 24, 26
Stockholm, 10, 84, 89
Stokes, John, 94
stoneware, salt glazed, 70, 92, 96, 99, 118, 133
Strachan, Shirley, 146, 165
Suez Canal, 62

Sumatra, 18, 109, 111
Sunda Strait, 18, 26, 111, 112
Surabaya, 112
Surat, 17, 153
survey techniques under water, 3, 6, 96, 116, 131, 141, 147
surveys, coastal, 21, 30, 31, 33, 51, 142, 145, 159
Swan River, 19
Swan River Colony, 101, 123, 127
Swan River Colony, 101, 127
Swedish warship, 89
Sydney, 13, 34, 36–8, 42, 46–8, 50–2, 55, 109, 150, 161
Sydney Bay, 36
Sydney Opera House, 48

Tahiti, 33, 38, 40, 44, 50, 134, 135
Tahitians and the Bounty mutiny, 38, 39, 142
Tamala Station, 96
Tasman, Abel, 31
Tasmania, 10–14, 36, 41, 42, 46, 47, 48, 50–2, 55, 62, 75, 80, 136, 142, 145, 146, 151, 153, 154, 155, 173
tea trade, 50
Terra Australis, 16, 28, 33
Texel, 23, 93, 94
theft at sea, 21
Thomas, D., 15n
Thomas, Nathaniel, 110
Thomas River, 34
Thomson, B., 144n
Thursday Island, 129
tides, 11, 98, 104
timber trade, 14, 59, 78
time capsules, 9, 31
Timor, 38, 51
Timor Sea, 12, 13
Tofua, 38, 66n
Table Bay, 18, 19
Tonga, 50
Torres Strait, 12, 17, 26, 28, 38, 39, 46, 47, 48, 52, 62, 142
tourism, 79, 82
trade routes and practices, 9, 13, 14, 15, 17, 18–20, 23, 24, 31, 33, 34, 40, 44, 45, 47, 51, 53, 55, 59–61, 109, 150, 151, 153, 168, 169
Trial Rocks, 20, 21, 98, 99
training:
  for diving, 170, 171
  for maritime archaeology, 80, 170, 171
Traitors Island, Houtman Abrolhos, 22
transition from sail to steam, 59, 60, 61, 62
Truganini, 44
Tuscany, 133

UNESCO, 4n
University of Newcastle, 161
University of Western Australia, 73, 80

Van Eyck, Els, 32n
Vavua Island, 135
Venus, 33
Verschoor, Jan, 17
vessel rigs:
  barque, 45, 52, 55, 58, 102, 121, 142, 154, 156, 158, 159, 173, 174-7
  brig, 34, 45, 47, 51, 52, 55, 114, 115, 118, 145, 150, 174-8
  brigantine, 45, 114, 159, 176
  barquentine, 156, 174
  cutter, 45, 153, 159, 175, 177
  dandy, 45
  ketch, 45
  schooner, 42, 45, 51, 52, 60, 112, 124, 142, 143, 156, 174-6, 178
  ship, 20, 41, 45, 52, 53, 55, 57-9, 109, 113, 121, 148, 150, 153, 154, 158, 177, 178
  sloop, 44, 45, 51
  smack, 45
  snow, 158
  yawl, 39
vessel types:
  canoe, 30
  clipper, 57, 58, 59, 156, 157, 173, 174
  corvette, 28
  cruiser, 174
  dredge, 176
  frigate, 26, 135
  galliot, 26
  hulk, 34
  hooker, 26
  lifeboat, 177
  lighter, 28
  liner, 64
  longboat, 20, 21, 26, 28, 42, 44
  lugger, 11, 62, 63, 124, 169, 174
  monitor, 62
  oil rig, 65
  pinnace, 17
  Pleistocene boat, 12, 14
  prahu, 30, 31
  skiff, 20, 21
  slaver, 11, 47, 102, 114, 117, 118, 148, 150, 174
  submarine, 174
  transport, 34, 38, 47, 48, 49, 50, 150
  tug, 173
  warship, 10, 34, 37, 44, 47, 62, 64, 89, 132, 133-5, 142, 162, 174
  yacht, 178
vessels:
  Australian-built, 11, 44, 51, 160, 169, 172
  India-built, 11, 153
vessels built of:
  composite materials, 57, 58, 159, 177
  iron, 57, 58, 59, 62, 124, 142, 154, 158, 159, 161, 162, 173, 174-8
  steel, 57, 58, 59, 159, 176
  wood, 57, 58, 156, 173-8
Vickery, Libby, 66n, 165n
Victoria, 13, 27, 28, 52, 55, 57, 58, 60, 62, 77, 80, 154-7, 174, 178
Victoria Archaeological Survey (VAS), 54, 155, 156, 160
  Maritime Archaeological Unit (MAU), 156
Vietnam, 32n
Viking ships, 109
de Vlamingh, Willem, 26
de Vries, Andries, 80

Waitangi Treaty House, New Zealand, 58
Wallabi Group, Houtman Abrolhos, 22, 80
Walters, —, 165n
Wandjina paintings, 30
war:
  at sea, 23, 30, 34, 40, 84, 168
  of 1812, 109, 110
  World War II, 62, 65, 162
Warrnambool, 27, 154, 156
Weser River, 89
West End, Fremantle, 75
West Indies, 33, 38, 47
Western Australia, 4, 10, 11, 13, 23, 28, 55, 77
Western Australia Company, 55
Western Australian Government:
  Conservation and Environment, Department of, 121
  Public Works Department, 120
  Tourist Development Authority, 79
Western Australian Institute of Technology, 80, 168, 170
Western Australian Maritime Museum, 75, 86, 90, 167
Western Australian Museum, 2, 4, 5, 63-5, 70-72, 75, 79, 80, 89, 91, 92, 92, 94-9, 102, 103, 109, 113, 118-21, 124, 127-9, 171
  Maritime Archaeology Advisory Committee, 80, 120
  Maritime Archaeology, Department of, 118, 127
  Materials Conservation and Restoration, Department of, 80, 89, 90
  Trustees, Board of, 79, 80, 82, 85
Westernport Bay, 155
Whale World Museum, Albany, 120
whale-chasers, steam, 120, 126, 176
whaling industry, 13, 28, 38, 47, 52, 53, 55, 78, 102-5, 121, 124, 169, 176
  bay whaling, 52, 124, 127
wheat trade, 50, 59
whetstones, 70
Whipple, A. B. C., 66n
Whitlam, Gough, 76
Whydah, 114
Williams, John, 37
Williamstown, 62
Wilson, Stan, 100n

Wilsons Promontory, 52, 159, 160
Wonnerup Estuary, 28
wood, 73, 80, 87–90, 98, 99, 109, 120, 131, 133, 138, 139, 145, 148, 150, 151, 153, 159
Wooddang, —, 113
Woodman Point, 115
wool trade, 58, 59, 60
Wright, Guy, 66n

Wysvliet, Marinus, 24

Yassi Ada, 5.
York, 79
Yorkshire, 53

Zuiderbaan, Lous, 121

Index of Vessels

[date in parentheses indicates shipwreck]

Aagtekerke (1726), 27
Aarhus (1894), 56, 142, 173
Abielle (see *Lively*)
SS *Admella* (1859), 56, 173
African (1863), 56, 175
Agincourt (1882), 56, 175
Amsterdam, 19
Amsterdam (1748), 95
Apollo (1827), 145
Arcadia (1900), 59
SS *Archimedes*, 60
Arpenteur (1849), 56, 174
Arthur, 150
Asterope (1883), 173

PS *Ballina* (1879), 56, 161, 162, 173
Batavia (1629), 9, 10, 21, 22, 24, 56, 69, 71-4, 78, 80-4, 86-91, 94, 95, 134, 168, 171, 174
Batoe Bassi (1880), 56, 176
HMS *Beagle*, 94
Belinda (1824), 52
Belle of Bunbury (1886), 56, 175
Ben Ledi (1879), 56, 174
Berwick (see *Sirius*)
Black Swan (1829), 52
HMS *Bounty* (1790), 38, 39, 41, 44, 134, 135, 142
Bremer Kogge (1380), 89
Britannia (1806), 52
Browse Island wreck (1870s), 56, 175

Calliance (1864), 178
Cambridgeshire (1875), 173
Campbell Macquarie (1812), 52
Carlisle Castle (1899), 56, 175
Cataraqui (1845), 53
Centaur (1874), 56, 174
HMS *Cerberus*, 62
PS *Ceres* (1836), 60
Cervantes (1844), 56, 174
Chalmers (1874), 56, 174
Charles, 20
Charles, 52

Charles W. Morgan, 52
Chaudiere (1883), 56, 175
SS *Cheynes 3*, 120, 126
Children (1839), 56, 156, 174
City of Adelaide (1854), 177
City of Edinburgh (1840), 173
SS *City of Launceston* (1865), 156, 157, 178
City of York (1899), 56, 175
SS *Clan Ranald* (1909), 56, 159, 173
Clarence (1850), 156, 157, 178
PS *Clonmel* (1841), 60, 156, 174
PS *Commodore* (1931), 162
USS *Constitution*, 135
Contest (1874), 56, 78, 176
Crown of England (1912), 56, 176
Cumberland (1830), 56, 102, 121, 122, 153, 176
Cutty Sark, 57

Dato (1893), 56, 175
Day Dawn (1890), 52, 121, 175
Denton Holme (1890), 56, 175
HMAS *Diamantina*, 128
Diana (1878), 56, 175
Don Francisco (see *James Matthews*)
Dordrecht, 19
HMS *Dryad*, 104
Duke of Bedford (1852), 59
Duyfken, 17, 19

Echo (1820), 52
SS *Eddystone* (1894), 56, 175
Edward Bonaventure, 19
Edwin Fox, 49, 50
Eendracht, 19
Eglinton (1852), 56, 102, 123-124, 168, 174
Eliza (1797), 42, 44
Elizabeth (1839), 56, 102, 174
Elizabeth Henrietta (1825), 51
Emden (1914), 56, 65, 163, 174
Emily Taylor (1830), 153
HMAS *Encounter*, 28
HMS *Endeavour* (1770), 33, 34, 56, 134, 142, 173
Endeavour (1795), 150, 153
L'Enterprise (1803), 51

Estramina (1816), 51
Europa (1897), 56, 175
EWS, 63
Eyre wreck (—1830), 56, 175

Fairy Queen (1875), 56, 174
Fame (1817), 48
SS *Fin* (1923), 56, 176
MV *Flamingo Bay*, 136
Foam (1893), 56, 142, 143, 173
Fortuyn (1727), 27
Francis (1805), 42, 43, 51

Geelvinck, 26
Geltwood (1876), 56, 173
Gem (1876), 56, 175
General Greene, 109, 112, 113
Geographe, 28
George (1830), 52
George III (1835), 48
SS *Georgette* (1876), 56, 175
Gilt Dragon (see *Vergulde Draeck*)
Gothenburg (1875), 56, 142, 173
Governor Ready (1829), 48
Graveland, 27
Grecian (1850), 55, 138, 159, 177
HMB *Griffon*, 114
HMS *Guardian* (1790), 37, 38, 40, 41

Hadda (1877), 56, 82, 102, 176
Harrington, 51
Helena Mena, 58
MV *Henrietta*, 82-4, 88
Henry (1825), 48
Hero of the Nile (1876), 56, 175
Highland Forest (1901), 176
Hunter, 113

I-124 (1942), 56, 65, 163, 174
Ida, 113
Iron King (1873), 177

James (1830), 56, 174
James Matthews (1841), 47, 56, 102, 114-8, 148-50, 168, 174
James Service (1878), 56, 175
Janet (1887), 56, 175
Jenny, 110
TSS *John Penn* (1879), 56, 161, 173

SS *Karrakatta* (1901), 56, 176
Katinka (1900), 56, 176
Kockenge, 24
SS *Koombana* (1912), 56, 176
Kormoran (1941), 65

La Bella (1905), 56, 156, 174
Lady Elizabeth (1878), 56, 175

Lady Juliana, 38
Lady Lyttleton (1867), 56, 102, 174
Lady Nelson (1825), 51
Lancier (1839), 56, 102, 174
Leeuwin, 19
Lightning (1869), 57
Litherland (1853), 173
Lively (c. 1810), 56, 104, 176
Loch Ard (1878), 56, 58, 154, 156, 174
Loch Vennachar (1905), 58, 158, 159, 173
Loda (1886), 59
SS *Lubra* (1898), 56, 176
MV *Lumen*, 129, 130

SS *Macedon* (1883), 56, 175
Madagascar (1853), 59
Mahogany ship, 27, 28, 154
Manfred (1879), 59, 175
Margaret Brock (1852), 56, 173
Marion (1851), 177
Marten (1878), 56, 175
Mary Rose (1545), 10, 89
Mauritius, 19
Mayflower (1880), 56, 176
Mayhill (1895), 56, 176
Meridian, 113
Mermaid (1829), 51, 56, 142, 173
Mersey (1805), 153
PS *Mimosa* (1863), 162, 173
Mira Flores (1886), 56, 175
Miranda (1846), 159, 160
Montebello (1906), 56, 159, 173
Monumental City (1853), 56, 60, 61, 156, 174
Morning Star (1814), 56, 142, 173
Mountain Maid (1856), 156, 178

Nashwauk (1855), 159, 177
MV *Nellie Melba*, 103
Neptune (1793), 150
Neva (1835), 49
New Zealander (1853), 155
Nijptangh, 26
Nobbys Head wreck (1931?), 173
Norfolk (1800), 51
Norma (1907), 177

Ocean Queen (1840), 56, 174
Ocean Ranger (1980—), 65
Omeo (1894), 56, 175

HMS *Pandora* (1791), 9, 10, 38-40, 44, 56, 78, 128-42, 153, 166, 173
SS *Penola*, 156
TSS *Pericles* (1910), 94
SS *Perth* (1887), 56, 175
PS *Phoenix* (1852), 60
Polly Woodside, 154, 155
Priestman dredge (1893), 56, 176
Prince Regent (1827), 48

INDEX OF VESSELS

Queen, 156
SS Quetta (1890), 56, 62, 142, 173

Rapid (1811), 47, 78, 102, 105, 106, 108, 110, 112–4, 148–50, 168, 175
Raven (1891), 56, 175
Ridderschap van Holland (1694), 26
Ringbolt Bay wreck (c. 1880), 56, 176
Rockingham (1830), 55
Rowley Shoals wreck (see Lively)
Royal Charlotte (1825), 48
Royal Exchange, 19
Royal Shepherd (1890), 161

San Miguel (1865), 177
Santiago (1945), 159, 177
Sardam, 23
SS Satara (1910), 64
Schah (1837), 118
Schiedam, 20
Schomberg (1855), 56, 57, 156, 174
PS Scottish Prince (1887), 142, 173
Sepia (1898), 56, 175
Shenandoah, 62
HMS Sirius (1790), 34–9, 44, 51, 56, 163–5, 167, 174
Sloepie, 26
Solway (1837), 159, 177
PS Sophia Jane, 60
South Australian (1837), 159
Star (1880), 56, 102, 175
Star of Greece (1888), 59, 158, 177
SS Sunbeam (1892), 62, 178
Sunset Beach wreck (see African)
HMS Supply, 36, 51
HMAS Sydney (1941), 65
Sydney Cove (1797), 41–4, 51, 56, 145–54, 173

SS Taupo (1881), 58

Thistle (1836), 153, 156
Thomas Nye (see Day Dawn)
Three Bees (1814), 48
Tigress (1848), 158, 159, 177
Topaz, 38
Trial (1622), 2, 19–21, 56, 76, 80, 82, 98, 99, 174
Tubal Caine (1862), 58

Ulidia (1893), 56, 175
Union, 52
Uribes (1942), 56, 176

Valetta (1897), 153, 175
Vergulde Draeck (1656), 11, 23, 24, 56, 69–73, 78, 80, 90–4, 120, 168, 174
HMS Victory, 135
Villalta (1897), 56, 176

Waeckende Boey, 24, 69
Wapen van Amsterdam (1619), 26
Wasa (1628), 10, 83, 89, 90, 135
Water Witch (1842), 159, 177
Weseltje, 26
William and Ann, 52
William Salthouse (1841), 53–55, 156, 157, 178
SS Windsor (1908), 56, 176

SS Xantho (1872), 63, 102, 124, 125, 161, 178

Yarra (1884), 176
SS Yongala (1911), 56, 64, 142, 144, 173

Zanoni (1867), 159, 177
Zedora (1875), 56, 174
Zeewijk (1727), 24, 26, 27, 56, 80, 94, 96, 124, 174
Zeewolf, 19, 94
SS Zuir (1902), 56, 175
Zuytdorp (1711), 24, 25, 56, 78, 80, 96–8, 174